D1573493

Encyclopaedia of Mathematical Sciences
Volume 121

Operator Algebras and Non-Commutative Geometry II

Subseries Editors:
Joachim Cuntz   Vaughan F.R. Jones

Springer
*Berlin
Heidelberg
New York
Hong Kong
London
Milan
Paris
Tokyo*

Joachim Cuntz
Georges Skandalis
Boris Tsygan

Cyclic Homology in Non-Commutative Geometry

Springer

Joachim Cuntz
University of Münster
Institute of Mathematics
Einsteinstraße 62
48149 Münster, Germany
e-mail: cuntz@math.uni-muenster.de

Georges Skandalis
Centre de Mathématiques de Jussieu
Université Paris 7 Denis Diderot
175 rue du Chevaleret
75013 Paris, France
e-mail: skandal@math.jussieu.fr

Boris Tsygan
Northwestern University
Department of Mathematics
2033 North Sheridan Avenue
Evanston, IL, 60208-2730 USA
e-mail: tsygan@math.northwestern.edu

Founding editor of the Encyclopaedia of Mathematical Sciences:
R. V. Gamkrelidze

Mathematics Subject Classification (2000): 16E40, 58J20, 19K56

ISSN 0938-0396
ISBN 3-540-40469-4 Springer-Verlag Berlin Heidelberg New York

This work is subject to copyright. All rights are reserved, whether the whole or part of the material is concerned, specifically the rights of translation, reprinting, reuse of illustrations, recitation, broadcasting, reproduction on microfilm or in any other way, and storage in data banks. Duplication of this publication or parts thereof is permitted only under the provisions of the German Copyright Law of September 9, 1965, in its current version, and permission for use must always be obtained from Springer-Verlag. Violations are liable for prosecution under the German Copyright Law.

Springer-Verlag Berlin Heidelberg New York
Springer-Verlag is a part of Springer Science+Business Media
springeronline.com
© Springer-Verlag Berlin Heidelberg 2004
Printed in Germany

The use of general descriptive names, registered names, trademarks, etc. in this publication does not imply, even in the absence of a specific statement, that such names are exempt from the relevant protective laws and regulations and therefore free for general use.

Typeset by LE-TEX Jelonek, Schmidt & Vöckler GbR, Leipzig
Cover Design: E. Kirchner, Heidelberg, Germany
Printed on acid-free paper   46/3142 db   5 4 3 2 1 0

Preface to the Encyclopaedia Subseries on Operator Algebras and Non-Commutative Geometry

The theory of von Neumann algebras was initiated in a series of papers by Murray and von Neumann in the 1930's and 1940's. A von Neumann algebra is a self-adjoint unital subalgebra M of the algebra of bounded operators of a Hilbert space which is closed in the weak operator topology. According to von Neumann's bicommutant theorem, M is closed in the weak operator topology if and only if it is equal to the commutant of its commutant. A *factor* is a von Neumann algebra with trivial centre and the work of Murray and von Neumann contained a reduction of all von Neumann algebras to factors and a classification of factors into types I, II and III.

C^*-algebras are self-adjoint operator algebras on Hilbert space which are closed in the norm topology. Their study was begun in the work of Gelfand and Naimark who showed that such algebras can be characterized abstractly as involutive Banach algebras, satisfying an algebraic relation connecting the norm and the involution. They also obtained the fundamental result that a commutative unital C^*-algebra is isomorphic to the algebra of complex valued continuous functions on a compact space – its spectrum.

Since then the subject of operator algebras has evolved into a huge mathematical endeavour interacting with almost every branch of mathematics and several areas of theoretical physics.

Up into the sixties much of the work on C^*-algebras was centered around representation theory and the study of C^*-algebras of type I (these algebras are characterized by the fact that they have a well behaved representation theory). Finite dimensional C^*-algebras are easily seen to be just direct sums of matrix algebras. However, by taking algebras which are closures in norm of finite dimensional algebras one obtains already a rich class of C^*-algebras – the so-called AF-algebras – which are not of type I. The idea of taking the closure of an inductive limit of finite-dimensional algebras had already appeared in the work of Murray-von Neumann who used it to construct a fundamental example of a factor of type II – the "hyperfinite" (nowadays also called approximately finite dimensional) factor.

One key to an understanding of the class of AF-algebras turned out to be K-theory. The techniques of K-theory, along with its dual, Ext-theory, also found immediate applications in the study of many new examples of C^*-algebras that arose in the end of the seventies. These examples include for instance "the noncommutative tori" or other crossed products of abelian C^*-algebras by groups of homeomorphisms and abstract C^*-algebras generated by isometries with certain relations, now known as the algebras $\mathcal{O}_n$. At the same time, examples of algebras were increasingly studied that codify data from differential geometry or from topological dynamical systems.

On the other hand, a little earlier in the seventies, the theory of von Neumann algebras underwent a vigorous growth after the discovery of a natural infinite family of pairwise nonisomorphic factors of type III and the advent of Tomita-Takesaki theory. This development culminated in Connes' great classification theorems for approximately finite dimensional ("injective") von Neumann algebras.

Perhaps the most significant area in which operator algebras have been used is mathematical physics, especially in quantum statistical mechanics and in the foundations of quantum field theory. Von Neumann explicitly mentioned quantum theory as one of his motivations for developing the theory of rings of operators and his foresight was confirmed in the algebraic quantum field theory proposed by Haag and Kastler. In this theory a von Neumann algebra is associated with each region of space-time, obeying certain axioms. The inductive limit of these von Neumann algebras is a C^*-algebra which contains a lot of information on the quantum field theory in question. This point of view was particularly successful in the analysis of superselection sectors.

In 1980 the subject of operator algebras was entirely covered in a single big three weeks meeting in Kingston Ontario. This meeting served as a review of the classification theorems for von Neumann algebras and the success of K-theory as a tool in C^*-algebras. But the meeting also contained a preview of what was to be an explosive growth in the field. The study of the von Neumann algebra of a foliation was being developed in the far more precise C^*-framework which would lead to index theorems for foliations incorporating techniques and ideas from many branches of mathematics hitherto unconnected with operator algebras.

Many of the new developments began in the decade following the Kingston meeting. On the C^*-side was Kasparov's KK-theory – the bivariant form of K-theory for which operator algebraic methods are absolutely essential. Cyclic cohomology was discovered through an analysis of the fine structure of extensions of C^*-algebras These ideas and many others were integrated into Connes' vast *Noncommutative Geometry* program. In cyclic theory and in connection with many other aspects of noncommutative geometry, the need for going beyond the class of C^*-algebras became apparent. Thanks to recent progress, both on the cyclic homology side as well as on the K-theory side, there is now a well developed bivariant K-theory and cyclic theory for a natural class of topological algebras as well as a bivariant character taking K-theory to cyclic theory. The 1990's also saw huge progress in the classification theory of nuclear C^*-algebras in terms of K-theoretic invariants, based on new insight into the structure of exact C^*-algebras.

On the von Neumann algebra side, the study of subfactors began in 1982 with the definition of the *index* of a subfactor in terms of the Murray-von Neumann theory and a result showing that the index was surprisingly restricted in its possible values. A rich theory was developed refining and clarifying the index. Surprising connections with knot theory, statistical mechanics and quantum field theory have been found. The superselection theory mentioned above turned out to have fascinating links to subfactor theory. The subfactors themselves were constructed in the representation theory of loop groups.

Beginning in the early 1980's Voiculescu initiated the theory of free probability and showed how to understand the free group von Neumann algebras in terms of random matrices, leading to the extraordinary result that the von Neumann algebra M of the free group on infinitely many generators has full fundamental group, i.e. pMp is isomorphic to M for every non-zero projection $p \in M$. The subsequent introduction of free entropy led to the solution of more old problems in von Neumann algebras such as the lack of a Cartan subalgebra in the free group von Neumann algebras.

Many of the topics mentioned in the (obviously incomplete) list above have become large industries in their own right. So it is clear that a conference like the one in Kingston is no longer possible. Nevertheless the subject does retain a certain unity and sense of identity so we felt it appropriate and useful to create a series of encylopaedia volumes documenting the fundamentals of the theory and defining the current state of the subject.

In particular, our series will include volumes treating the essential technical results of C^*-algebra theory and von Neumann algebra theory including sections on noncommutative dynamical systems, entropy and derivations. It will include an account of K-theory and bivariant K-theory with applications and in particular the index theorem for foliations. Another volume will be devoted to cyclic homology and bivariant K-theory for topological algebras with applications to index theorems. On the von Neumann algebra side, we plan volumes on the structure of subfactors and on free probability and free entropy. Another volume shall be dedicated to the connections between operator algebras and quantum field theory.

October 2001   subseries editors:
                                            Joachim Cuntz
                                            Vaughan Jones

Preface

Cyclic homology was introduced in the early eighties independently by Connes and Tsygan. They came from different directions. Connes wanted to associate homological invariants to K-homology classes and to describe the index pairing with K-theory in that way, while Tsygan was motivated by algebraic K-theory and Lie algebra cohomology. At the same time Karoubi had done work on characteristic classes that led him to study related structures, without however arriving at cyclic homology properly speaking.

Many of the principal properties of cyclic homology were already developed in the fundamental article of Connes and in the long paper by Feigin–Tsygan. In the sequel, cyclic homology was recognized quickly by many specialists as a new intriguing structure in homological algebra, with unusual features. In a first phase it was tried to treat this structure as well as possible within the traditional framework of homological algebra. The cyclic homology groups were computed in many examples and new important properties such as product structures, excision for H-unital ideals, or connections with cyclic objects and simplicial topology, were established. An excellent account of the state of the theory after that phase is given in the book of Loday.

This book is an attempt at an account of the present state of cyclic theory that covers in particular a number of more recent results that have changed the face of the theory to a certain extent. An essential feature of cyclic homology which is not so well captured by the ordinary graded cyclic homology groups HC_n is the S-operator, which reflects in fact a filtration – rather than a grading – on cyclic homology, along with the associated periodic cyclic theory introduced by Connes. The most characteristic properties of cyclic homology hold only after stabilization by the S-operator. Important steps beyond the classical graded framework were taken by Connes when he introduced the so-called entire theory, and by Goodwillie in his result on nilpotent extensions. The spirit here is to work not with the individual cyclic homology groups in finite dimension, but with the infinite cyclic bicomplex. This global approach also turned out to be the most appropriate for various index theorems. It was put on a new basis by Cuntz–Quillen who used free (or more generally quasi-free) resolutions of an algebra and a non-commutative de Rham complex, to describe all the cyclic-type invariants associated with that algebra. This also led to proofs of excision for general extensions in periodic, entire or other cyclic theories. Excision in its bivariant form in turn immediately leads to a natural multiplicative transformation – the bivariant Chern–Connes character – from bivariant K-theory, provided that this can be defined, to bivariant cyclic homology. This general transformation contains most previously constructed characters from K-theoretic invariants to cyclic theory invariants. The formalism using the non-commutative de Rham complex and free resolutions of an algebra is also the natural basis for cyclic homology theories generalizing the entire theory that can be applied to a wide class of topological algebras, including Banach- and C^*-algebras.

Other new techniques in cyclic homology theory were developed by Nest and Tsygan who systematically treat cyclic chains as noncommutative differential forms and Hochschild cochains as noncommutative multivector fields. In this way they recover the standard algebra arising in calculus to a surprisingly large extent. The resulting algebraic structures on the Hochschild and cyclic complexes are applied to prove far-reaching generalizations of the Atiyah–Singer index theorem culminating in the theorem of Bressler–Nest–Tsygan. The algebraic structures were further enriched in the work of Tamarkin and Tamarkin–Tsygan and used to reprove and extend Kontsevich's results on classification of deformation quantizations.

On the other hand, from the beginning, Connes developed cyclic theory into a powerful tool in non-commutative geometry. Striking applications of cyclic homology to geometry appeared already in his work on the transverse fundamental class for actions of discrete groups on a manifold and for transversally oriented foliations, including an interpretation of the Godbillon–Vey class in cyclic cohomology. Another such application is the proof of the Novikov conjecture for hyperbolic groups by Connes–Moscovici using cyclic cohomology and a 'higher' index theorem. More recently, Connes–Moscovici proved two other important index theorems which required new techniques in cyclic homology. The first one is a general abstract local index formula for spectral triples, which works for general algebras sharing some of the characteristic features of the algebra of pseudodifferential operators, and applies to many natural operators in non-commutative geometry. Locality is reflected by the fact that one uses only the Dixmier–Wodzicki trace for the description of the cyclic cocycles determining the index. The application of their local index formula to tranversally hypoelliptic operators on foliations led Connes and Moscovici to the introduction of a new type of cyclic homology for Hopf algebras. The total index in transverse geometry is expressed in terms of a characteristic map from the cyclic homology of a Hopf algebra playing the role of a symmetry group in this situation. The contribution by G. Skandalis in this volume is devoted to an outline of these last two theorems. A much more complete account is planned for another volume in this series.

Joachim Cuntz and Boris Tsygan

Contents

Cyclic Theory, Bivariant K-Theory and the Bivariant Chern-Connes Character
Joachim Cuntz 1

Cyclic Homology
Boris Tsygan 73

Noncommutative Geometry, the Transverse Signature Operator, and Hopf Algebras [after A. Connes and H. Moscovici]
Georges Skandalis (translated by Raphaël Ponge and Nick Wright) 115

Index 135

List of Contributors

Joachim Cuntz
Mathematisches Institut
Universität Münster
Einsteinstr. 62
48149 Münster
Germany
cuntz@math.uni-muenster.de

Boris Tsygan
Department of Mathematics
Northwestern University
Evanston, IL, 60208
USA
tsygan@@math.nwu.edu

Georges Skandalis
Université de Paris VII
U.F.R. de Mathématiques
Case Postale 7012
2, Place Jussieu
75251 Paris Cedex 05
France

Cyclic Theory, Bivariant K-Theory and the Bivariant Chern-Connes Character[*]

Joachim Cuntz

1	**Introduction**	2
2	**Cyclic Theory**	5
2.1	Preliminaries	5
2.2	Cyclic homology via the cyclic bicomplex	9
2.3	Operators on differential forms	13
2.4	The periodic theory	16
2.5	Cyclic homology via the X-complex	21
2.6	Cyclic homology as non-commutative de Rham theory	27
2.7	Homotopy invariance for cyclic theory	29
2.8	Morita invariance	31
2.9	Excision	33
2.10	Chern character for K-theory elements	34
3	**Cyclic Theory for locally convex algebras**	36
3.1	General modifications	36
3.2	De Rham theory for differentiable manifolds	38
3.3	Cyclic homology for Schatten ideals	38
3.4	Cyclic cocycles associated with Fredholm modules	40
4	**Bivariant K-Theory**	42
4.1	Bivariant K-theory for locally convex algebras	43
4.2	The bivariant Chern-Connes character	48
5	**Infinite-dimensional Cyclic Theories**	51
5.1	Entire cyclic cohomology	51
5.2	Local cyclic cohomology	57

---

[*] Research supported by the Deutsche Forschungsgemeinschaft

A	Locally convex algebras	61
A.1	Algebras of differentiable functions	62
A.2	The smooth tensor algebra	62
A.3	The free product of two m-algebras	63
A.4	The algebra of smooth compact operators	64
A.5	The Schatten ideals $\ell^p(H)$	65
A.6	The smooth Toeplitz algebra	65
B	Standard extensions	66
B.1	The suspension extension	66
B.2	The free extension	67
B.3	The universal two-fold trivial extension	67
B.4	The Toeplitz extension	68
References		69

1 Introduction

The two fundamental "machines" of non-commutative geometry are (bivariant) topological K-theory and cyclic homology. In the present contribution we describe these two theories and their connections. Cyclic theory can be viewed as a far reaching generalization of the classical de Rham cohomology, while bivariant K-theory includes the topological K-theory of Atiyah-Hirzebruch as a special case.

The classical commutative theories can be extended to a degree of generality which is quite striking. It is important to note however that this extension is by no means simply based on generalizations of the existing classical methods. The constructions are quite different and give, in the commutative case, a new approach and an unexpected interpretation of the well-known classical theories. One aspect is that some of the properties of the two theories become visible only in the non-commutative category. For instance, both theories have certain universality properties in this setting.

Bivariant K-theory has first been defined and developed by Kasparov on the category of C^*-algebras (possibly with the action of a locally compact group) thereby unifying and decisively extending previous work by Atiyah-Hirzebruch, Brown-Douglas-Fillmore and others. Kasparov also applied his bivariant theory to obtain striking positive results on the Novikov conjecture. Very recently [13], it was discovered that in fact, bivariant topological K-theories can be defined on a wide variety of topological algebras ranging from rather general locally convex algebras to e.g. Banach algebras or C^*-algebras (in fact, even algebras without a specified topology can be covered to some extent). If E is the covariant functor from such a category C of algebras given by topological K-theory or also by periodic cyclic homology, then it possesses the following three fundamental properties:

(E1) E is diffotopy invariant, i.e., the evaluation map ev_t in any point $t \in [0,1]$ induces an isomorphism $E(ev_t) : E(\mathcal{C}^\infty([0,1], \mathcal{A})) \to E(\mathcal{A})$ for any $\mathcal{A}$ in C. Here $\mathcal{C}^\infty([0,1], \mathcal{A})$ denotes the algebra of $\mathcal{A}$-valued C^∞-functions on $[0,1]$.

(E2) E is stable, i.e., the canonical inclusion $\iota : \mathcal{A} \to \mathcal{K} \hat{\otimes} \mathcal{A}$, where $\mathcal{K}$ denotes the algebra of infinite $\mathbb{N} \times \mathbb{N}$-matrices with rapidly decreasing coefficients, induces an isomorphism $E(\iota)$ for any $\mathcal{A}$ in C.

(E3) E is half-exact, i.e., each extension $0 \to \mathcal{I} \to \mathcal{A} \to \mathcal{B} \to 0$ in C admitting a continuous linear splitting induces a short exact sequence $E(\mathcal{I}) \to E(\mathcal{A}) \to E(\mathcal{B})$

"Diffotopy", i.e., differentiable homotopy is used in (E1) for technical reasons in connection with the homotopy invariance properties of cyclic homology. For K-theory, it could also be replaced by ordinary continuous homotopy.

It turns out that the bivariant K-functor is the universal functor from the given category C of algebras into an additive category D (i.e., the morphism sets $D(\mathcal{A}, \mathcal{B})$ are abelian groups) that has these three properties, see 4.15 below.

Cyclic theory is a homology theory that has been discovered, starting from K-theory, independently by Connes and Tsygan, [4] (announced in 1981), [50]. Connes's construction was motivated by Kasparov's formalism for K-homology and by the desire to construct a target for a noncommutative Chern character, see also [27]. A crucial role is played by so-called Fredholm modules or spectral triples. Also Tsygan's work is closely related to K-theory, [50]. In fact, in his approach, the new theory was originally called "additive K-theory" and he pointed out that it is an additive version of Quillen's definition of algebraic K-theory. It was immediately realized that cyclic homology has intimate connections with de Rham theory, Lie algebra homology, group cohomology and index theorems.

The theory with the really good properties is the periodic theory introduced by Connes. Periodic cyclic homology HP_* satisfies the three properties (E1), (E2), (E3), [4, 17, 20, 53]. Combining this fact with the universality property of bivariant K-theory leads to a multiplicative transformation (the bivariant Chern-Connes character) from bivariant K-theory to bivariant periodic cyclic theory. This transformation is a vast generalization of the classical Chern character in differential geometry.

The principal aim of this contribution is to give an account of cyclic theory. Here, everything works in parallel algebraically as well as for locally convex algebras (such as the algebra of C^∞-functions on a smooth manifold or of algebras of pseudodifferential operators, to name just two examples). Cyclic theory can be introduced using rather different complexes, each one of them having its own special virtues. Specifically, we will use the following complexes or bicomplexes:

- The cyclic bicomplex with various realizations:

$$CC^n(A) \quad \text{with boundary operators} \quad b, b', Q, 1-\lambda$$
$$\overline{CC}^n(\widetilde{A}) \quad \text{with boundary operator} \quad B-b$$
$$\Omega(A) \quad \text{with boundary operator} \quad B-b$$

  The cyclic bicomplex is well suited for the periodic theory as well as for the $\mathbb{Z}$-graded ordinary theory and for the connections between both.
- The Connes complex C_λ^n

  It has the advantage, that concrete *finite-dimensional* cocycles often arise naturally as elements of C_λ^n. The connection with Hochschild cohomology also fits naturally into this picture.
- The X-complex of any complete quasi-free extension of the given algebra A. This complex is very useful for a conceptual explanation of the properties of periodic theory, in particular for proving excision, for the connections with topological K-theory and the bivariant Chern-Connes character. It also is the natural framework for all infinite-dimensional versions of cyclic homology (analytic and entire as well as asymptotic and local theory).

In the first chapter we discuss the basic properties of cyclic theory. We don't try to give a complete account of all aspects of cyclic theory. The reader is referred to the contributions by Tsygan and Skandalis in this same volume for complementary accounts, treating in particular applications to index theorems, and to [33] for an exposition with emphasis on the aspects of cyclic theory related to classical homological algebra. We focus on those notions and results which are most closely related to topological K-theory. This includes homotopy invariance, Morita invariance, excision but also explicit formulas for the Chern character associated to idempotents and invertibles.

In the following chapter we describe the (more or less obvious) provisions that have to be made in cyclic theory for locally convex algebras. An important example is given by the Schatten ideals. We determine their cyclic cohomology and discuss the cyclic cocycles associated by a Chern character transformation to finitely summable Fredholm modules.

In Chap. 3 we turn to a description of bivariant K-theory and to the construction of the bivariant Chern-Connes character, which generalizes the Chern character for idempotents, invertibles or Fredholm modules mentioned before (as well as the classical Chern character of course).

As we pointed out already above, cyclic homology and (with some restrictions) bivariant K-theory can be defined on different categories of algebras - purely algebraically for algebras (over $\mathbb{R}$ or $\mathbb{C}$) or on categories of topological algebras such as locally convex algebras, Banach algebras or C^*-algebras. There are different variants of the two theories which are adapted to the different categories. In this text we treat cyclic theory in the purely algebraic case (however restricting to algebras over a field of characteristic 0) on the one hand. Concerning cyclic theory for topological algebras on the other hand, we have to make a choice. For the classical ("finite-dimensional") cyclic theories we concentrate on the category of what we call m-algebras. These are particularly nice locally convex algebras (projective limits of Banach algebras). Their advantage is that both, ordinary cylic homology and bivariant topological K-theory make perfect sense and that the bivariant Chern-Connes-character can be constructed nicely on this category. We note however that both theories described here, cyclic theory and bivariant K-theory still make sense for more general (though not arbitrary) topological algebras.

Moreover, besides the theories described in Chaps. 1–3, there are different versions of cyclic theory and of bivariant K-theory especially designed for other categories of topological algebras. In particular, for the categories of Banach algebras or C^*-algebras there are special variants of cyclic theory, namely the entire and the local cyclic theory, [5, 43]. They are best treated in the X-complex picture. An interesting new feature here is the existence of *infinite-dimensional* cohomology classes. These theories, as well as a bivariant character from Kasparov's KK-theory to the local cyclic theory, are discussed in Chap. 4 which has been written by Ralf Meyer.

2 Cyclic Theory

2.1 Preliminaries

Some constructions on algebras

We list in this section some examples of algebras and constructions on algebras that will be used later. For simplicity we always assume that the ground field is $\mathbb{C}$.

Adjoining a unit

Since it is important to do all constructions in the category of non-unital algebras we will frequently use the procedure of adjoining a unit to a given algebra A.

Definition 2.1. *The unitization $\tilde{A}$ of an algebra A is the vector space $\mathbb{C} \oplus A$ with multiplication given by*

$$(\lambda, x)(\mu, y) = (\lambda\mu, \lambda y + \mu x + xy)$$

$\tilde{A}$ is a unital algebra and contains A as an ideal of codimension 1.

Algebras of polynomial functions

For any algebra A, we can form the algebra $A[t]$ of polynomials with coefficients in A.

Two homomorphisms $\alpha, \beta : A \to B$ of algebras are called polynomially homotopic, if there is a homomorphism $\varphi : A \to B[t]$ such that $\varphi(x)(0) = \alpha(x), \varphi(x)(1) = \beta(x)$ for $x \in A$. Polynomial homotopy is not an equivalence relation, in general.

We denote by $A[z, z^{-1}]$ or by $A(z)$ the algebra of Laurent polynomials with coefficients in A.

The tensor algebra

Given any vector space V, we define the (non-unital) tensor algebra over V as
$$TV = V \oplus V \otimes V \oplus V^{\otimes^3} \oplus \ldots$$
TV is an algebra with the product given by concatenation of tensors. We denote by $\sigma : V \to TV$ the linear map, that maps V to the first summand in TV. The map σ has the following universal property: Let $s : V \to \mathcal{A}$ be an arbitrary linear map from V into an algebra $\mathcal{A}$. Then there is a unique algebra homomorphism $\tau_s : TV \to \mathcal{A}$ such that $\tau_s \circ \sigma = s$.

The tensor algebra is polynomially contractible, i.e., the identity map of TV is polynomially homotopic to 0. A polynomial homotopy $\varphi_t : TV \to TV$, such that $\varphi_0 = 0, \varphi_1 = \mathrm{id}$, is given by $\varphi_t = \tau_{t\sigma}, t \in [0, 1]$.

The free product of two algebras

Let A and B be algebras. The algebraic free product (in the non-unital category) of A and B is the following algebra
$$A * B = A \oplus B \oplus (A \otimes B) \oplus (B \otimes A) \oplus (A \otimes B \otimes A) \oplus \ldots$$
This direct sum is taken over all tensor products where the factors A and B alternate. The multiplication is given by concatenation of tensors but where the multiplication $A \otimes A \to A$ and $B \otimes B \to B$ is used to simplify all terms where two elements in A or two elements in B meet.

The canonical inclusions $\iota_1 : A \to A * B$ and $\iota_2 : B \to A * B$ have the following universal property: Let $\alpha : A \to E$ and $\beta : B \to E$ be two homomorphisms into an algebra E. Then there is a unique homomorphism $\alpha * \beta : A * B \to E$, such that $(\alpha * \beta) \circ \iota_1 = \alpha$ and $(\alpha * \beta) \circ \iota_2 = \beta$.

The algebra of finite matrices of arbitrary size

For any algebra A, we can form the algebra $M_\infty(A)$ consisting of matrices $(a_{ij})_{i,j \in \mathbb{N}}$ with $a_{ij} \in A$, $i, j = 0, 1, 2 \ldots$, and only finitely many a_{ij} non-zero. Equivalently
$$M_\infty(A) = \varinjlim_n M_n(A) \tag{1}$$
where, as usual, $M_n(A)$ is embedded into $M_{n+1}(A)$ in the upper left corner.

Elementary homological algebra

A complex is a $\mathbb{Z}$-graded vector space C with an endomorphism ∂ of degree -1 such that $\partial^2 = 0$. The map ∂ is called the boundary operator of C. We will also use $\mathbb{Z}/2$-graded complexes ("supercomplexes"). They can be viewed as periodic ordinary complexes.

Definition 2.2. *The homology $H(C)$ of a complex C is the quotient space*

$$H(C) = \operatorname{Ker} \partial / \operatorname{Im} \partial$$

$H(C)$ *is a graded vector space with* $H_n(C) = (\operatorname{Ker} \partial \cap C_n)/\partial(C_{n+1})$.

A linear map between two complexes C and D is said to have degree k if it maps C_n to D_{n+k} for all n. The Hom-complex $\operatorname{Hom}(C, D)$ is the (graded) vector space consisting of linear maps φ from the vector space C to the vector space D which are linear combinations of such homogeneous maps. The boundary map on $\operatorname{Hom}(C, D)$ is given by the graded commutator

$$\partial(\varphi) = \partial_D \circ \varphi - (-1)^{\deg \varphi} \varphi \circ \partial_C$$

with the boundary maps. A map $\varphi : C \to D$ of degree 0 is called a morphism of complexes if $\partial \varphi = 0$, i.e., if it commutes with the boundary operators. Any such morphism induces a morphism

$$\varphi_* : H(C) \longrightarrow H(D)$$

Two morphisms in $\operatorname{Hom}(C, D)$ are said to be (chain) homotopic if they differ by a boundary in $\operatorname{Hom}(C, D)$. Homotopic morphisms induce the same maps on homology. Two complexes are called homotopy equivalent if there exist

$$\varphi \in \operatorname{Hom}(C, D), \ \partial \varphi = 0, \ \deg \varphi = 0$$

and

$$\psi \in \operatorname{Hom}(C, D), \ \partial \psi = 0, \ \deg \psi = 0$$

such that $\varphi\psi$ and $\psi\varphi$ are homotopic to id. Finally, the complex C will be said to be contractible if it is homotopy equivalent to 0 (this is the case iff there exists a map $s : C \to C$ of degree 1, such that $\operatorname{id}_C = \partial s + s\partial$). The homology of a contractible complex is 0.

A sequence

$$\cdots \xrightarrow{\alpha_{n+2}} G_{n+1} \xrightarrow{\alpha_{n+1}} G_n \xrightarrow{\alpha_n} G_{n-1} \xrightarrow{\alpha_{n-1}} \cdots$$

of morphisms between vector spaces or abelian groups is called exact, if, for each n, $\operatorname{Ker} \alpha_n = \operatorname{Im} \alpha_{n+1}$.

Proposition 2.3. *Let*
$$0 \longrightarrow C \xrightarrow{\iota} D \xrightarrow{\pi} E \longrightarrow 0$$
be an extension of complexes, i.e., an exact sequence of (graded) vector spaces, where the maps ι and π are morphisms of complexes. Then there is a map $\delta : H(E) \to H(C)$ of degree -1 and an exact sequence of the form

$$\begin{array}{ccc} & H(D) & \\ {}^{\iota_*}\nearrow & & \searrow^{\pi_*} \\ H(C) & \xleftarrow{\delta} & H(E) \end{array}$$

This can be written as a long exact sequence of the form

$$\xrightarrow{\pi_*} H_{n+1}E \xrightarrow{\delta} H_n C \xrightarrow{\iota_*} H_n D \xrightarrow{\pi_*} H_n E \xrightarrow{\delta} H_{n-1}C \xrightarrow{\iota_*} H_{n-1}D \xrightarrow{\pi_*}$$

Proof. Given $[x]$ in $H(E)$ with $\partial(x) = 0$, put $\delta([x]) = \partial(z)$, where $z \in D$ is such that $\pi(z) = x$.

Definition 2.4. *A bicomplex is a $\mathbb{Z} \times \mathbb{Z}$-graded vector space or, equivalently, a family $(C_{pq})_{p,q \in \mathbb{Z}}$ of vector spaces with two endomorphisms ∂_h and ∂_v of degree $(-1, 0)$ and $(0, -1)$, respectively, such that*

$$\partial_h^2 = \partial_v^2 = \partial_h \partial_v + \partial_v \partial_h = 0$$

∂_h *and* ∂_v *are called the horizontal and vertical boundary maps of C.*
The total complex for C is the graded vector space D whose component of degree n is given by the infinite product

$$D_n = \prod_{p+q=n} C_{pq}$$

(The letter D stands for diagonals). With the boundary operator $\partial = \partial_h + \partial_v$ it becomes a complex.

Remark 2.5. Instead of the direct product, one can also take the direct sum $D_n = \bigoplus_{p+q=n} C_{pq}$ in the definition of the total complex. Both variants will be used in the sequel.

For each p, the family $(C_{pq})_{q \in \mathbb{Z}}$ defines a complex with boundary operator ∂_v. We call its homology the vertical homology. The horizontal homology is defined similarly for each q. We will often consider bicomplexes where $C_{pq} = 0$ for $p < 0$ or $q < 0$. The following proposition is then often useful.

Proposition 2.6. *Let $C = C_{pq}$ and $E = E_{pq}$ be bicomplexes with $C_{pq} = 0$ and $E_{pq} = 0$ for $p < 0$ or $q < 0$, and let D and F denote the total complexes of C and E, respectively. Assume that $\alpha, \beta : C \to E$ are two morphisms of bicomplexes which, for each p, induce the same map on the vertical homology of C and E, respectively. Then the induced maps $\alpha_*, \beta_* : H(D) \to H(F)$ coincide.*

2.2 Cyclic homology via the cyclic bicomplex

In this section, we introduce cyclic homology using, what are probably the most fundamental operators in cyclic theory (b, b', λ, Q), and we explain the long exact sequence connecting cyclic homology with Hochschild homology. The important properties of cyclic theory will be discussed only later.

The cyclic bicomplex $CC_{pq}(A)$, $p, q \in \mathbb{N}$, for an algebra A is the following

$$
\begin{array}{ccccccc}
\downarrow & & \downarrow & & \downarrow & & \\
A^{\otimes 3} & \xleftarrow{1-\lambda} & A^{\otimes 3} & \xleftarrow{Q} & A^{\otimes 3} & \xleftarrow{1-\lambda} & \ldots \\
\downarrow b & & \downarrow -b' & & \downarrow b & & \\
A^{\otimes 2} & \xleftarrow{1-\lambda} & A^{\otimes 2} & \xleftarrow{Q} & A^{\otimes 2} & \xleftarrow{1-\lambda} & \ldots \\
\downarrow b & & \downarrow -b' & & \downarrow b & & \\
A & \xleftarrow{1-\lambda} & A & \xleftarrow{Q} & A & \xleftarrow{1-\lambda} & \ldots
\end{array}
$$

i.e.,
$$CC_{pq}(A) = A^{\otimes(q+1)} \qquad p, q \geq 0$$

In this section and later, bicomplexes are understood to continue with 0 outside the represented region (thus here $C_{pq} = 0$ for $p < 0$ or $q < 0$). The map b' is given by

$$b'(x_0 \otimes x_1 \otimes \ldots \otimes x_n) = \sum_{j=0}^{n-1} (-1)^j x_0 \otimes \ldots \otimes x_j x_{j+1} \otimes \ldots \otimes x_n$$

while b is the usual Hochschild operator

$$b(x_0 \otimes \ldots \otimes x_n) = b'(x_0 \otimes \ldots \otimes x_n) + (-1)^n x_n x_0 \otimes x_1 \otimes \ldots \otimes x_{n-1}$$

Finally, λ is Connes's signed cyclic permutation operator

$$\lambda(x_0 \otimes \ldots \otimes x_n) = (-1)^n x_n \otimes x_0 \otimes \ldots \otimes x_{n-1}$$

and $Q = \sum_{j=0}^n \lambda^j$ is averaging with respect to signed cyclic permutations, thus

$$Q(x_0 \otimes \ldots \otimes x_n) = \sum_{j=0}^n (-1)^{jn} x_j \otimes \ldots \otimes x_n \otimes x_0 \otimes \ldots \otimes x_{j-1}$$

One has $b^2 = 0$, $b'^2 = 0$ and $(1-\lambda)Q = Q(1-\lambda) = 0$. Moreover, $CC_{pq}(A)$ is in fact a bicomplex, due to the fundamental identities

$$Qb = b'Q$$
$$(1-\lambda)b' = b(1-\lambda)$$

The rows of the cyclic bicomplex $CC_{pq}(A)$ are obviously exact as long as $p > 0$.

Definition 2.7. $HC_n A$ is the homology of the total complex of the cyclic bicomplex $CC_{pq}(A)$, i.e., $HC_n A$ is the homology of the complex

$$\cdots \longrightarrow D_n \xrightarrow{\partial} D_{n-1} \xrightarrow{\partial} \cdots \xrightarrow{\partial} D_1 \xrightarrow{\partial} D_0 \longrightarrow 0$$

where
$$D_n = CC_{0,n} \oplus CC_{1,(n-1)} \oplus \ldots \oplus CC_{n,0}$$
$$= A^{\otimes(n+1)} \oplus A^{\otimes n} \oplus \ldots \oplus A$$

are the diagonals in $CC_{pq}(A)$ and ∂ is the sum of the respective boundary operators $b, b', 1 - \lambda, Q$.

Consider the bicomplex $CC_*^{\text{Hochschild}}$ consisting of the first two columns of CC_*:

$$\begin{array}{ccc} \downarrow & \downarrow & \\ A^{\otimes 3} & \xleftarrow{1-\lambda} & A^{\otimes 3} \\ \downarrow b & \downarrow -b' & \\ A^{\otimes 2} & \xleftarrow{1-\lambda} & A^{\otimes 2} \qquad (CC_*^{\text{Hochschild}})\\ \downarrow b & \downarrow -b' & \\ A & \xleftarrow{1-\lambda} & A \end{array}$$

and its total complex

$$\cdots \xrightarrow{\partial'} D_n^{\text{Hochschild}} \xrightarrow{\partial'} D_{n-1}^{\text{Hochschild}} \xrightarrow{\partial'} \cdots \xrightarrow{\partial'} D_0^{\text{Hochschild}} \longrightarrow 0$$

where
$$D_n^{\text{Hochschild}} = A^{\otimes(n+1)} \oplus A^{\otimes n}, \quad n \geq 1, \qquad D_0^{\text{Hochschild}} = A$$

$$\text{and} \quad \partial' = \begin{pmatrix} b & (1-\lambda) \\ 0 & -b' \end{pmatrix}$$

Definition 2.8. For any (possibly nonunital) algebra A, the Hochschild homology $HH_*(A)$ is defined as the homology of the complex

$$\cdots \xrightarrow{\partial'} D_n^{\text{Hochschild}} \xrightarrow{\partial'} D_{n-1}^{\text{Hochschild}} \xrightarrow{\partial'} \cdots \xrightarrow{\partial'} D_0^{\text{Hochschild}} \longrightarrow 0$$

If A has a unit, then the second column in $CC_*^{\text{Hochschild}}$ is obviously exact and can therefore be disregarded for the computation of $HH_*(A)$. We therefore find in this case the usual definition of the Hochschild homology of A (with coefficients in A), namely as the homology of the complex

$$\cdots \xrightarrow{b} A^{\otimes 3} \xrightarrow{b} A^{\otimes 2} \xrightarrow{b} A \longrightarrow 0$$

More generally, if this holds, we call A H-unital, see [53].

Definition 2.9. *The algebra A is called H-unital, if the second column in $CC_*^{\text{Hochschild}}$, i.e., the complex*

$$\cdots \xrightarrow{b'} A^{\otimes 3} \xrightarrow{b'} A^{\otimes 2} \xrightarrow{b'} A \longrightarrow 0$$

is exact.

For any H-unital algebra A, the Hochschild homology $HH(A)$ is the homology of the complex

$$\cdots \xrightarrow{b} A^{\otimes 3} \xrightarrow{b} A^{\otimes 2} \xrightarrow{b} A \longrightarrow 0$$

If A is unital, then the Hochschild homology of A has a well known interpretation as a derived functor. In particular it can be computed from any projective A-bimodule resolution of the A-bimodule A. This method applies in many cases.

Theorem 2.10. *There is a long exact sequence (Connes's SBI-sequence) of the form*

$$\longrightarrow HC_nA \xrightarrow{S} HC_{n-2}A \xrightarrow{B} HH_{n-1}A \xrightarrow{I} HC_{n-1}A \xrightarrow{S} HC_{n-3}A \xrightarrow{B}$$
$$\cdots \xrightarrow{I} HC_1A \xrightarrow{S} HC_{-1}A \longrightarrow HH_0A \xrightarrow{I} HC_0A \longrightarrow 0$$

Proof. This is the long exact sequence associated to the obvious exact sequence of complexes

$$0 \longrightarrow D_n^{\text{Hochschild}} \longrightarrow D_n \longrightarrow D_{n-2} \longrightarrow 0$$

The operators I and S in the above sequence are simply induced by the inclusion $D_n^{\text{Hochschild}} \to D_n$ and by the projection $D_n \to D_{n-2}$, while the operator B is given by the boundary map. Note that $HC_{-1}A = 0$ and that $I : HH_0A \xrightarrow{\cong} HC_0A$ is an isomorphism. If the Hochschild homology of A is known, then the SBI-sequence allows in many cases to determine the cyclic homology groups for A recursively.

Consider the following sub-bicomplex $CC_*^{\text{trivial}}(A)$ of $CC_{pq}(A)$

$$\begin{array}{ccccccc}
\downarrow & & \downarrow & & \downarrow & & \\
(1-\lambda)A^{\otimes 3} & \xleftarrow{1-\lambda} & A^{\otimes 3} & \xleftarrow{Q} & A^{\otimes 3} & \xleftarrow{1-\lambda} & \cdots \\
\downarrow b & & \downarrow -b' & & \downarrow b & & \\
(1-\lambda)A^{\otimes 2} & \xleftarrow{1-\lambda} & A^{\otimes 2} & \xleftarrow{Q} & A^{\otimes 2} & \xleftarrow{1-\lambda} & \cdots \\
\downarrow b & & \downarrow -b' & & \downarrow b & & \\
(1-\lambda)A & \xleftarrow{1-\lambda} & A & \xleftarrow{Q} & A & \xleftarrow{1-\lambda} & \cdots
\end{array} \quad (CC_*^{\text{trivial}}(A))$$

Since the rows in $CC_*^{\text{trivial}}(A)$ are exact, its total complex D^{trivial} has no homology. We denote by $C^\lambda(A)$ the quotient complex D/D^{trivial}. Explicitly

$$C_n^\lambda(A) = D_n/D_n^{\text{trivial}} = A^{\otimes(n+1)}/(1-\lambda)A^{\otimes(n+1)}$$

Since $H_n(D^{\text{trivial}}) = 0$, we immediately get

Proposition 2.11. *For any algebra A, the cyclic homology $HC_n(A)$ can be computed as the homology of the complex*

$$\cdots \longrightarrow C_n^\lambda(A) \xrightarrow{b} C_{n-1}^\lambda(A) \xrightarrow{b} \cdots \longrightarrow C_0^\lambda(A) \longrightarrow 0$$

This is Connes's original definition of cyclic homology.

The two-sided (or periodic) cyclic complex CC_*^{twosided} continues CC_* to the left. It is defined by $CC_{pq}^{\text{twosided}}(A) = A^{\otimes(q+1)}$ $p, q \in \mathbb{Z}, q \geq 0$ with boundary operators as follows

$$\begin{array}{ccccccc}
& \downarrow & & \downarrow & & \downarrow & \\
\cdots \xleftarrow{Q} & A^{\otimes 3} & \xleftarrow{1-\lambda} & A^{\otimes 3} & \xleftarrow{Q} & A^{\otimes 3} & \xleftarrow{1-\lambda} \cdots \\
& \downarrow b & & \downarrow -b' & & \downarrow b & \\
\cdots \xleftarrow{Q} & A^{\otimes 2} & \xleftarrow{1-\lambda} & A^{\otimes 2} & \xleftarrow{Q} & A^{\otimes 2} & \xleftarrow{1-\lambda} \cdots \\
& \downarrow b & & \downarrow -b' & & \downarrow b & \\
\cdots \xleftarrow{Q} & A & \xleftarrow{1-\lambda} & A & \xleftarrow{Q} & A & \xleftarrow{1-\lambda} \cdots \\
\end{array} \quad (CC_{pq}^{\text{twosided}}(A))$$

The left cyclic complex CC_*^{left} is the left-hand side of CC_*^{twosided}

$$\begin{array}{ccccccc}
& \downarrow & & \downarrow & & \downarrow & \\
\cdots \xleftarrow{1-\lambda} & A^{\otimes 3} & \xleftarrow{Q} & A^{\otimes 3} & \xleftarrow{1-\lambda} & A^{\otimes 3} & \\
& \downarrow -b' & & \downarrow b & & \downarrow -b' & \\
\cdots \xleftarrow{1-\lambda} & A^{\otimes 2} & \xleftarrow{Q} & A^{\otimes 2} & \xleftarrow{1-\lambda} & A^{\otimes 2} & \\
& \downarrow -b' & & \downarrow b & & \downarrow -b' & \\
\cdots \xleftarrow{1-\lambda} & A & \xleftarrow{Q} & A & \xleftarrow{1-\lambda} & A & \\
\end{array} \quad (CC_*^{\text{left}})$$

Consider the total complexes D^{left}, D^{twosided}, D of the bicomplexes

$$CC_{pq}^{\text{left}}(A), \quad CC_{pq}^{\text{twosided}}(A), \quad CC_{pq}(A)$$

respectively, where here we use direct sums rather than direct products for the definition of the diagonals in the total complex. We get an exact sequence of complexes

$$0 \longrightarrow D^{\text{left}} \longrightarrow D^{\text{twosided}} \longrightarrow D \longrightarrow 0$$

Since the rows of $CC_{pq}^{\text{twosided}}(A)$ are exact, a standard staircase argument shows that D^{twosided} has no homology (this argument would not work for the direct product total complex). Therefore, from the long exact sequence 2.3

$$HC_n(A) = H_n(D) \cong H_{n-1}(D^{\text{left}})$$

whence the following alternative description of cyclic homology

Proposition 2.12. *For any algebra A, the cyclic homology $HC_n(A)$ can be computed as*
$$HC_n(A) = H_{n-1}(D^{\text{left}})$$
where D^{left} is the total complex of $CC_{pq}^{\text{left}}(A)$.

By using the dual complexes, the analogue of any statement in this section is also valid for cohomology in place of homology.

In particular, cyclic cohomology can be defined using the dual $C_\lambda^n A = (C_n^\lambda A)'$ of $C_n^\lambda A$, i.e.,

$$C_\lambda^n A = \left\{ f \;\middle|\; \begin{array}{l} f \text{ is an } (n+1)\text{-linear functional on } A^{n+1} \text{ such that} \\ f(x_n, x_0, x_1, \ldots, x_{n-1}) = (-1)^n f(x_0, x_1, \ldots, x_n) \end{array} \right\}$$

The boundary operator on $C_\lambda^n A$ is the restriction of the Hochschild operator b.

A **cyclic n-cocycle** then is an $(n+1)$-linear functional such that

$$f(x_n, x_0, x_1, \ldots, x_{n-1}) = (-1)^n f(x_0, x_1, \ldots, x_n)$$

and

$$f(x_0 x_1, x_2, \ldots, x_{n+1}) - f(x_0, x_1 x_2, \ldots, x_{n+1}) + \ldots$$
$$+ (-1)^n f(x_0, \ldots, x_n x_{n+1}) + (-1)^{n+1} f(x_{n+1} x_0, x_1, \ldots, x_n) = 0$$

This is the original definition of a cyclic cocycle due to A. Connes. It also is the form in which cyclic cocycles appear often in practice.

2.3 The algebra ΩA of abstract differential forms over A and its operators

The boundary operators in the cyclic bicomplex take a simpler form and can be "explained" if interpreted as operators on the algebra ΩA of abstract differential forms over A. Another deeper and even more conceptual explanation will be presented in Sect. 2.21.

Given an algebra A, we denote by ΩA the universal algebra generated by $x \in A$ with relations of A and symbols $dx, x \in A$, where dx is linear in x and satisfies $d(xy) = xd(y) + d(x)y$. We do not impose $d1 = 0$, i.e., if A has a unit, $d1 \neq 0$. ΩA is a direct sum of subspaces $\Omega^n A$ generated by linear combinations of $x_0 dx_1 \ldots dx_n$, and $dx_1 \ldots dx_n$, $x_j \in A$. This decomposition makes ΩA into a graded algebra. As usual, we write $\deg(\omega) = n$ if $\omega \in \Omega^n A$.

As a vector space, for $n \geq 1$,
$$\Omega^n A \cong \tilde{A} \otimes A^{\otimes n} \cong A^{\otimes(n+1)} \oplus A^{\otimes n} \tag{2}$$

(where $\tilde{A}$ is A with a unit adjoined, cf. 2.1, and $1 \otimes x_1 \otimes \cdots \otimes x_n$ corresponds to $dx_1 \ldots dx_n$). The operator d is defined on ΩA by

$$d(x_0 dx_1 \ldots dx_n) = dx_0 dx_1 \ldots dx_n$$
$$d(dx_1 \ldots dx_n) = 0 \tag{3}$$

The operator b is defined by

$$b(\omega dx) = (-1)^{\deg \omega}[\omega, x]$$
$$b(dx) = 0, \, b(x) = 0, \qquad x \in A, \, \omega \in \Omega A \tag{4}$$

Then clearly $d^2 = 0$ and one easily computes that also $b^2 = 0$. Under the isomorphism in equation (2) d becomes

$$d(x_0 \otimes \ldots \otimes x_n) = 1 \otimes x_0 \otimes \ldots \otimes x_n \quad , \text{ if } x_0 \in A$$
$$d(1 \otimes x_1 \otimes \ldots \otimes x_n) = 0$$

while b corresponds to the usual Hochschild operator

$$b(\widetilde{x_0} \otimes x_1 \otimes \ldots \otimes x_n) =$$
$$\widetilde{x_0} x_1 \otimes \ldots \otimes x_n + \sum_{j=2}^n (-1)^{j-1} \widetilde{x_0} \otimes \ldots \otimes x_{j-1} x_j \otimes \ldots \otimes x_n$$
$$+ (-1)^n x_n \widetilde{x_0} \otimes x_1 \otimes \ldots \otimes x_{n-1}, \qquad \widetilde{x_0} \in \tilde{A}, \, x_1, \ldots, x_n \in A$$

Another important natural operator is the degree (or number) operator:
$$N(\omega) = \deg(\omega) \omega \tag{5}$$

We also define the Karoubi operator κ on ΩA by
$$\kappa = 1 - (db + bd) \tag{6}$$

Explicitly, κ is given by
$$\kappa(\omega dx) = (-1)^{\deg \omega} dx \, \omega$$

The operator κ satisfies $(\kappa^n - 1)(\kappa^{n+1} - 1) = 0$ on Ω^n. Therefore, by linear algebra, there is a projection operator P on ΩA corresponding to the generalized eigenspace for 1 of the operator $\kappa = 1 - (db + bd)$.

Lemma 2.13. *Let $L = (Nd)b + b(Nd)$. Then $\Omega A = \operatorname{Ker} L \oplus \operatorname{Im} L$ and P is exactly the projection onto $\operatorname{Ker} L$ in this splittting.*

Proof. This follows from the identity

$$L = (\kappa - 1)^2(\kappa^{n-1} + 2\kappa^{n-2} + 3\kappa^{n-3} + \ldots + (n-1)\kappa + n)$$

The operator L thus behaves like a "selfadjoint" operator. It can be viewed as an abstract Laplace operator on the algebra of abstract differential forms ΩA. The elements in the image of P are then "abstract harmonic forms".

By construction, P commutes with b, d, N. Thus setting $B = NPd$ one finds $Bb + bB = PL = 0$ and $B^2 = 0$.

Explicitly, B is given on $\omega \in \Omega^n$ by the formula

$$B(\omega) = \sum_{j=0}^{n} \kappa^j d\omega$$

Under the isomorphism in equation (2), this corresponds to:

$$B(x_0 dx_1 \ldots dx_n) = dx_0 dx_1 \ldots dx_n + (-1)^n dx_n dx_0 \ldots dx_{n-1}$$
$$+ \ldots + (-1)^{nn} dx_1 \ldots dx_n dx_0$$

The preceding identities show that we obtain a bicomplex - the (B,b)-bicomplex - in the following way

$$\begin{array}{ccccccc}
\downarrow b & & \downarrow b & & \downarrow b & & \downarrow b \\
\Omega^3 A & \xleftarrow{B} & \Omega^2 A & \xleftarrow{B} & \Omega^1 A & \xleftarrow{B} & \Omega^0 A \\
\downarrow b & & \downarrow b & & \downarrow b & & \\
\Omega^2 A & \xleftarrow{B} & \Omega^1 A & \xleftarrow{B} & \Omega^0 A & & \\
\downarrow b & & \downarrow b & & & & \\
\Omega^1 A & \xleftarrow{B} & \Omega^0 A & & & & \\
\downarrow b & & & & & & \\
\Omega^0 A & & & & & &
\end{array} \quad (7)$$

One can rewrite the (B,b)-bicomplex (7) using the isomorphism $\Omega^n A \cong A^{\otimes(n+1)} \oplus A^{\otimes n}$ in equation (2). An easy computation shows that the operator $b : \Omega^n A \to \Omega^{n-1} A$ corresponds under this isomorphism to the operator $A^{\otimes(n+1)} \oplus A^{\otimes n} \to A^{\otimes n} \oplus A^{\otimes(n-1)}$ which is given by the matrix

$$\begin{pmatrix} b & (1-\lambda) \\ 0 & -b' \end{pmatrix}$$

where b', b and λ are the operators discussed in Sect. 2.2.

Similarly, the operator $B : \Omega^n A \to \Omega^{n+1} A$ corresponds to the operator $A^{\otimes(n+1)} \oplus A^{\otimes n} \to A^{\oplus(n+2)} \oplus A^{\otimes(n+1)}$ given by the 2 × 2-matrix

$$\begin{pmatrix} 0 & 0 \\ Q & 0 \end{pmatrix}$$

where Q is as in Sect. 2.2.

This shows that the (B,b)-bicomplex is just another way of writing the cyclic bicomplex, i.e., the total complex D^Ω for the (B,b)-bicomplex is exactly isomorphic to the total complex D for the cyclic bicomplex. As a consequence we have

Proposition 2.14. $HC_n A$ is the homology of the complex

$$\longrightarrow D_n^\Omega \xrightarrow{B'-b} D_{n-1}^\Omega \xrightarrow{B'-b} \ldots \xrightarrow{B'-b} D_1^\Omega \xrightarrow{B'-b} D_0^\Omega \longrightarrow 0$$

where

$$D_{2n}^\Omega = \Omega^0 A \oplus \Omega^2 A \oplus \ldots \oplus \Omega^{2n} A$$

$$D_{2n+1}^\Omega = \Omega^1 A \oplus \Omega^3 A \oplus \ldots \oplus \Omega^{2n+1} A$$

and B' is the truncated B-operator, i.e., $B' = B$ on the components $\Omega^k A$ of D_n^Ω, except on the highest component $\Omega^n A$, where it is 0.

$HH_n(A)$ is the homology of the complex

$$\longrightarrow \Omega^n A \xrightarrow{b} \Omega^{n-1} A \xrightarrow{b} \ldots \xrightarrow{b} \Omega^1 A \xrightarrow{b} \Omega^0 A \longrightarrow 0$$

Remark 2.15. Assume that A has a unit 1. We may introduce in ΩA the additional relation $d(1) = 0$, i.e., divide by the ideal M generated by $d(1)$ (this is equivalent to introducing the relation $1 \cdot \omega = \omega$ for all ω in ΩA). We denote the quotient by $\overline{\Omega}A$. Now M is a graded subspace, invariant under b and B, and its homology with respect to b is trivial. The preceding proposition is thus still valid if we use $\overline{\Omega}A$ in place of ΩA. The convention $d(1) = 0$ is often used (implicitly) in the literature. In some cases it simplifies computations considerably. There are however situations where one cannot reduce the computations to the unital situation (in particular this is true for the excision problem).

2.4 The periodic theory

The periodic theory is the one that has the really good properties like homotopy invariance, Morita invariance and excision. It generalizes the classical de Rham theory to the non-commutative setting.

Periodic cyclic homology

Let A be an algebra. We denote by $\widehat{\Omega}A$ the infinite product

$$\widehat{\Omega}A = \prod_n \Omega^n A$$

and by $\widehat{\Omega}^{ev}A$, $\widehat{\Omega}^{odd}A$ its even and odd part, respectively. $\widehat{\Omega}A$ may be viewed as the (periodic) total complex for the bicomplex

$$
\begin{array}{ccccccc}
& \downarrow{-b} & & \downarrow{-b} & & \downarrow{-b} & & \downarrow{-b} \\
\longleftarrow & \Omega^3 A & \xleftarrow{B} & \Omega^2 A & \xleftarrow{B} & \Omega^1 A & \xleftarrow{B} & \Omega^0 A \\
& \downarrow{-b} & & \downarrow{-b} & & \downarrow{-b} & \\
\longleftarrow & \Omega^2 A & \xleftarrow{B} & \Omega^1 A & \xleftarrow{B} & \Omega^0 A & \\
& \downarrow{-b} & & \downarrow{-b} & & & \\
\longleftarrow & \Omega^1 A & \xleftarrow{B} & \Omega^0 A & & & \\
& \downarrow{-b} & & & & & \\
\longleftarrow & \Omega^0 A & & & & &
\end{array}
$$

Similarly, the (continuous for the filtration topology) dual $(\widehat{\Omega}A)'$ of $\widehat{\Omega}A$ is

$$(\widehat{\Omega}A)' = \bigoplus_n (\Omega^n A)'$$

Definition 2.16. *The periodic cyclic homology $HP_*(A)$, $* = 0, 1$, is defined as the homology of the $\mathbb{Z}/2$-graded complex*

$$\widehat{\Omega}^{ev} A \underset{B-b}{\overset{B-b}{\rightleftarrows}} \widehat{\Omega}^{odd} A$$

and the periodic cyclic cohomology $HP^*(A)$, $* = 0, 1$, is defined as the homology of the $\mathbb{Z}/2$-graded complex

$$(\widehat{\Omega}^{ev} A)' \underset{B-b}{\overset{B-b}{\rightleftarrows}} (\widehat{\Omega}^{odd} A)'$$

Now by definition, S is the projection

$$D_n^{\Omega} = \Omega^n \oplus \Omega^{n-2} \oplus \ldots \quad \longrightarrow \quad D_{n-2}^{\Omega} = \Omega^{n-2} \oplus \Omega^{n-4} \oplus \ldots$$

where D_n^{Ω} is as in 2.14. Therefore we get

$$\widehat{\Omega}A = \varprojlim_S (D_{2n}^{\Omega} \oplus D_{2n+1}^{\Omega})$$

and

$$(\widehat{\Omega}A)' = \varinjlim_{S'} (D_{2n}^{\Omega} \oplus D_{2n+1}^{\Omega})'$$

We deduce

Proposition 2.17. *We still denote by S the maps $HC_n \to HC_{n-2}$ and $HC^n \to HC^{n+2}$ induced by the operator defined above. For any algebra A and $* = 0, 1$ one has*
$$HP^*(A) = \varinjlim_S HC^{2n+*}A$$
and an exact sequence
$$0 \to \varprojlim_S{}^1 HC_{2n+*+1}A \to HP_*A \to \varprojlim_S HC_{2n+*}A \to 0$$
(where as usual $\varprojlim_S{}^1 HC_{2n+*+1}A$ is defined as
$$\left(\prod_n HC_{2n+*+1}A\right) \Big/ (1-s)\left(\prod_n HC_{2n+*+1}A\right)$$
s being the shift on the infinite product).

The bivariant theory

Now $\widehat{\Omega}A$ is in a natural way a complete metric space (with the metric induced by the filtration on ΩA—the distance of families (x_n) and (y_n) in $\prod \Omega^n A$ is $\leq 2^{-k}$ if the k first x_i and y_i agree). We call this the filtration topology and denote by $\mathrm{Hom}\,(\widehat{\Omega}A, \widehat{\Omega}B)$ the set of continous linear maps $\widehat{\Omega}A \to \widehat{\Omega}B$. It can also be described as
$$\mathrm{Hom}\,(\widehat{\Omega}A, \widehat{\Omega}B) = \varprojlim_m \varinjlim_n \mathrm{Hom}\left(\bigoplus_{i\leq n} \Omega^i A, \bigoplus_{j\leq m} \Omega^j B\right).$$
It is a $\mathbb{Z}/2$-graded complex with boundary map
$$\partial \varphi = \varphi \circ \partial - (-1)^{\deg(\varphi)} \partial \circ \varphi$$
where $\partial = B - b$.

Definition 2.18. *Let A and B be algebras. Then the bivariant periodic cyclic homology $HP_*(A, B)$ is defined as the homology of the Hom-complex*
$$HP_*(A, B) = H_*(\mathrm{Hom}\,(\widehat{\Omega}A, \widehat{\Omega}B)) \quad * = 0, 1$$

It is not difficult to see that the $\mathbb{Z}/2$-graded complex $\widehat{\Omega}\mathbb{C}$ is (continuously with respect to the filtration topology) homotopy equivalent to the trivial complex
$$\mathbb{C} \rightleftarrows 0$$
Therefore
$$HP_*(\mathbb{C}, B) = HP_*(B) \quad \text{and} \quad HP_*(A, \mathbb{C}) = HP^*(A)$$

There is an obvious product

$$HP_i(A_1, A_2) \times HP_j(A_2, A_3) \to HP_{i+j}(A_1, A_3)$$

induced by the composition of elements in $\operatorname{Hom}(\widehat{\Omega}A_1, \widehat{\Omega}A_2)$ and $\operatorname{Hom}(\widehat{\Omega}A_2, \widehat{\Omega}A_3)$, which we denote by $(x, y) \mapsto x \cdot y$. In particular, $HP_0(A, A)$ is a unital ring with unit 1_A given by the identity map on $\widehat{\Omega}A$.

An element $\alpha \in HP_*(A, B)$ is called invertible if there exists $\beta \in HP_*(B, A)$ such that $\alpha \cdot \beta = 1_A \in HP_0(A, A)$ and $\beta \cdot \alpha = 1_B \in HP_0(B, B)$. An invertible element of degree 0, i.e., in $HP_0(A, B)$ will also be called an HP-equivalence. Such an HP-equivalence exists in $HP_0(A, B)$ if and only if the supercomplexes $\widehat{\Omega}A$ and $\widehat{\Omega}B$ are continuously homotopy equivalent. Multiplication by an invertible element α on the left or on the right induces natural isomorphisms $HP_*(B, D) \cong HP_*(A, D)$ and $HP_*(D, A) \cong HP_*(D, B)$ for any algebra D.

Remark 2.19. One can also define a $\mathbb{Z}$-graded version of the bivariant cyclic theory, [26], as follows. Say that a linear map $\alpha : \widehat{\Omega}A \to \widehat{\Omega}B$, continuous for the filtration topology, is of order $\leq k$ if $\alpha(F^n \widehat{\Omega}A) \subset F^{n-k} \widehat{\Omega}B$ for $n \geq k$, where $F^n \widehat{\Omega}A$ is the infinite product of $b(\Omega^{n+1})$, Ω^{n+1}, Ω^{n+2}, ... We denote by $\operatorname{Hom}^k(\widehat{\Omega}A, \widehat{\Omega}B)$ the set of all maps of order $\leq k$. This is, for each k, a subcomplex of the $\mathbb{Z}/2$-graded complex $\operatorname{Hom}(\widehat{\Omega}A, \widehat{\Omega}B)$. We can define

$$HC_n(A, B) = H_i(\operatorname{Hom}^n(\widehat{\Omega}A, \widehat{\Omega}B)) \text{ where } i \in \{0, 1\}, i \equiv n \bmod 2$$

The bivariant theory $HC_n(A, B)$ has a product $HC_n(A, B) \times HC_m(B, C) \to HC_{n+m}(A, C)$ and satisfies

$$HC_n(\mathbb{C}, B) = HC_n(B) \quad \text{and} \quad HC_n(A, \mathbb{C}) = HC^n(A)$$

In general, there exist elements in $HP_*(A, B)$ which are not in the range of the natural map $HC_{2n+*}(A, B) \to HP_*(A, B)$ for any n, [16].

Mixed complexes.

We present still another construction of cyclic theory which is sometimes convenient for explicit computations of cyclic homology in terms of Hochschild homology.

Following [29], a *mixed complex* is a $\mathbb{Z}$-graded complex $\bigoplus_n M_n$, bounded below, whose differential (usually denoted b) has degree -1, which is equipped with an operator B of degree $+1$ satisfying $[b, B] = B^2 = 0$. The homology of M as a complex, i.e., with respect to b, is by definition the *Hochschild homology* of M:

$$HH_n M = H_n M$$

Associated to M is the $\mathbb{Z}/2$-graded complex given by

$$\widehat{M} = \prod_n M_n$$

equipped with the usual even-odd grading and the differential $B - b$. The *periodic cyclic homology* is then the homology of this supercomplex:

$$HP_\nu M = H_\nu \widehat{M}, \quad \nu \in \mathbb{Z}/2$$

The *cyclic homology* of M is defined as

$$HC_n M = H_n(D)$$

where D is the complex given by

$$D_n = \bigoplus_{p \geq 0} M_{n-2p}$$

with boundary operator $B' - b$, where $B' : M_n \to M_{n-1}$ is, as in 2.14, the truncated B-operator.

We have seen that, for any algebra A, there is a natural mixed complex, given by $(\Omega(A), b, B)$ and the Hochschild, cyclic and periodic cyclic homology of A is by definition the Hochschild, cyclic and periodic cyclic homology of this mixed complex.

The following simple observation is useful for computations of cyclic homology invariants.

Proposition 2.20. *Let $\varphi : (M, b, B) \to (M', b, B)$ be a morphism of mixed complexes (i.e., φ maps M_n to M'_n for all n and φ commutes with the b- and B-operators).*

If φ is a quasi-isomorphism from (M, b) to (M', b), then φ induces isomorphisms on the cyclic and periodic cyclic homology of M and M'.

Proof. This follows from the analogue of 2.10 and 2.17 for mixed complexes.

For instance, the cyclic homology of smooth (in the sense of Grothendieck) algebras, of the locally convex algebra of smooth functions on a compact differentiable manifold or of the irrational rotation algebra (noncommutative torus) can be computed that way. In these cases, the Hochschild homology $HH(A)$ can be computed and there is an operator $\dot{B}$ of degree 1 on $HH(A)$ as well as a surjective map of mixed complexes

$$(\Omega(A), b, B) \to (HH(A), 0, \dot{B})$$

see [4, 34]. In the case of a smooth algebra or of the smooth functions on a compact manifold, $HH(A)$ is simply the algebra of ordinary classical differential forms and the operator $\dot{B}$ is the classical de Rham operator d.

2.5 Cyclic homology via the X-complex

The simple X-complex has been introduced first by Quillen [46] in connection with cyclic homology for coalgebras. It has been applied to free or quasi-free resolutions of a given algebra in [15–17]. This yields a radically different approach to cyclic theory which is particularly useful for the study of periodic theories and cyclic theories for topological algebras, for proving the excision theorem, and for understanding the connection with topological K-theory.

The X-complex

Definition 2.21. *Let A be an algebra. The X-complex $X(A)$ is the $\mathbb{Z}/2$-graded complex*

$$A \underset{b}{\overset{\natural d}{\rightleftarrows}} \Omega^1 A_\natural$$

where $\Omega^1 A_\natural = \Omega^1 A/b(\Omega^2 A) = \Omega^1 A/[A, \Omega^1 A]$ and $\natural : \Omega^1 A \to \Omega^1 A_\natural$ is the canonical quotient map.

Recall that, from the definition of b, its image exactly consists of all commutators, thus $b(\Omega^2 A) = [A, \Omega^1 A]$.

The X-complex is constructed from the lowest level of the (B, b)-bicomplex

$$\begin{array}{ccccc}
\downarrow b & & \downarrow b & & \downarrow b \\
\longleftarrow \Omega^2 A & \overset{B}{\longleftarrow} & \boxed{\Omega^1 A} & \overset{B=d}{\longleftarrow} & \boxed{\Omega^0 A} \\
\downarrow b & & \downarrow b & & \\
\longleftarrow \boxed{\Omega^1 A} & \overset{B=d}{\longleftarrow} & \boxed{\Omega^0 A} & & \\
\downarrow b & & & & \\
\longleftarrow \boxed{\Omega^0 A} & & & &
\end{array}$$

To be precise, the X-complex is the total complex of the quotient of the bicomplex

$$\begin{array}{ccccccc}
\downarrow b & & \downarrow b & & \downarrow b & & \downarrow b \\
\longleftarrow \Omega^3 A & \overset{B}{\longleftarrow} & \Omega^2 A & \overset{B}{\longleftarrow} & \Omega^1 A & \overset{B}{\longleftarrow} & \Omega^0 A \\
\downarrow b & & \downarrow b & & \downarrow b & & \\
\longleftarrow \Omega^2 A & \overset{B}{\longleftarrow} & \Omega^1 A & \overset{B}{\longleftarrow} & \Omega^0 A & & \\
\downarrow b & & \downarrow b & & & & \\
\longleftarrow \Omega^1 A & \overset{B}{\longleftarrow} & \Omega^0 A & & & & \\
\downarrow b & & & & & & \\
\longleftarrow \Omega^0 A & & & & & &
\end{array}$$

by the sub-bicomplex

$$
\begin{array}{ccccc}
\downarrow b & & \downarrow b & & \downarrow b \\
\longleftarrow \Omega^3 A & \xleftarrow{B} & \Omega^2 A & \xleftarrow{B} & b(\Omega^2 A) \\
\downarrow b & & \downarrow b & & \\
\longleftarrow \Omega^2 A & \xleftarrow{B} & b(\Omega^2 A) & & \\
\downarrow b & & & & \\
\longleftarrow b(\Omega^2 A) & & & &
\end{array}
\qquad (8)
$$

where $\Omega^0 A$ is replaced by 0 and $\Omega^1 A$ is replaced by $b(\Omega^2 A)$. The X-complex is periodic of period two.

Quasi-free algebras

The X-complex is specifically designed to compute the periodic cyclic theory for a special class of "smooth" algebras - the quasi-free algebras, because for quasi-free algebras the subcomplex by which we divide is contractible.

Proposition-Definition 2.22 *([15, 48]) Let A be an algebra. The following conditions are equivalent:*

(1) *Let $0 \to N \to S \xrightarrow{q} B \to 0$ be an extension of algebras where the ideal N is nilpotent (i.e., $N^k = \{0\}$ for some $k \geq 1$) and $A \xrightarrow{\alpha} B$ a homomorphism. Then there exists a homomorphism $A \xrightarrow{\alpha'} S$ such that $q \circ \alpha' = \alpha$.*
(2) *A has cohomological dimension ≤ 1 with respect to Hochschild cohomology.*
(3) *$\Omega^1 A$ is a projective bimodule over A.*

The algebra A is called <u>quasi-free</u> if these equivalent conditions are satisfied.

The list of equivalent conditions characterizing quasi-freeness can be extended. In particular, (2) is equivalent to

(2)' There exists a linear map $\varphi : A \to \Omega^2 A$ such that

$$\varphi(xy) = x\varphi(y) + \varphi(x)y - dxdy \qquad (9)$$

(i.e., the universal Hochschild 2-cocycle $c : A \times A \to \Omega^2 A$ defined by $c(x, y) = dxdy$ is the boundary of φ: $c = b\varphi$).

and (3) is equivalent to

(3)' There exists a linear map $\nabla : \Omega^1 A \to \Omega^2 A$ (a "right connection") satisfying
$$\nabla(a\omega) = a\nabla\omega$$
$$\nabla(\omega a) = (\nabla\omega)a + \omega da \qquad (a \in A, \omega \in \Omega^2 A)$$

The equivalence between (2)' and (3)' is based on the correspondence between φ and ∇ given by the equality $\nabla(xdy) = x\varphi(y)$. For explicit computations the property (9) of quasi-free algebras often is particularly useful.

Proposition 2.23. *Let A be quasi-free and let $HX_*(A), * = 0, 1$ be the homology of the complex $X(A)$. Then $HX_*(A) = HP_*A$.*

Proof. The proof is based on a contraction of the infinite product $(b\Omega^2 A, \Omega^2 A, \Omega^3 A, \ldots)$ inside $\widehat{\Omega}A$ obtained from the connection ∇. Or, equivalently on showing that the total complex of the sub-bicomplex (8) is contractible.

The most important examples of quasi-free algebras are free algebras. In particular, for any vector space V, the (non-unital) tensor algebra

$$TV = V \oplus V^{\otimes 2} \oplus V^{\otimes 3} \oplus \ldots$$

is quasi-free. It obviously satisfies condition (1) in 2.22. In fact, it admits liftings in any (not necessarily nilpotent) extension. The map φ of (9) is determined by the condition that $\varphi(x) = 0$ for all generators $x \in V$. If $\mathcal{A}$ is an m-algebra, then the tensor algebra $T\mathcal{A}$ in the category of m-algebras is quasi-free. Other examples of quasifree algebras include $\mathbb{C}$, $M_n(\mathbb{C})$ or $\mathbb{C}[z, z^{-1}]$. One obtains from this

Corollary 2.24.

$$HP_0(\mathbb{C}) = \mathbb{C} \qquad HP_1(\mathbb{C}) = 0$$
$$HP_0(\mathbb{C}[z, z^{-1}]) = \mathbb{C} \ HP_1(\mathbb{C}[z, z^{-1}]) = \mathbb{C}$$

Connection between the X-complex for a free resolution and the $B - b$-bicomplex

We will now see that, using a natural filtration, the X-complex for the tensor algebra over a given algebra A miraculously leads to the $B - b$-complex.

Definition 2.25. *Let T be an algebra and J an ideal in T. The J-adic filtration on $X(T)$ is defined by*

$$F^{2n+1}X(T) = J^{n+1} \oplus \natural(J^{n+1}dT + J^n dJ)$$
$$F^{2n}X(T) = (J^{n+1} + [J^n, T]) \oplus \natural(J^n dT)$$

Remark 2.26. This defines a decreasing filtration which is closely related to the inverse system $X(T/J^n)$. In fact, let F^n be the J-adic filtration of $X(T)$ as above and G^n the kernel of the quotient map $X(T) \to X(T/J^n)$. Then

$$G^{n+1} \subset F^{2n+1} \subset F^{2n} \subset G^n \subset \ldots$$

Therefore $\widehat{X}(T) = X(\widehat{T})$ where

$$\widehat{T} = \varprojlim_n T/J^n \qquad \text{and} \qquad \widehat{X}(T) = \varprojlim_n X(T)/F^n$$

We are now going to apply this to $T = TA$ and $J = JA$ where TA is the non-unital tensor algebra over A and JA is the ideal in the standard free extension $0 \to JA \to TA \to A \to 0$ of A (see Appendix B.2). The connection between the X-complex for free algebras and the cyclic bicomplex is based on the following results.

Proposition 2.27. ([16]) *Let A be an algebra.*

(1) The following defines an associative product (Fedosov product) on ΩA:

$$\omega_1 \circ \omega_2 = \omega_1 \omega_2 + (-1)^{\deg(\omega_1)} d\omega_1 d\omega_2$$

where $\omega_1 \omega_2$ is the ordinary product of ω_1, ω_2 in ΩA. Clearly, the even forms $\Omega^{ev} A$ form a subalgebra of $(\Omega A, \circ)$ for $\circ$.

(2) Let $p : A \to TA$ be the canonical linear inclusion of A into the tensor algebra over A and, for $x, y \in A$, $\omega(x, y) = p(xy) - pxpy$. Then the map

$$\alpha : x_0 dx_1 dx_2 \ldots dx_{2n-1} dx_{x_{2n}} \mapsto x_0 \omega(x_1, x_2) \ldots \omega(x_{2n-1}, x_{2n})$$

defines an <u>isomorphism</u> of algebras $(\Omega^{ev} A, \circ) \xrightarrow{\alpha} TA$.

(3) The map $\omega dx \mapsto \natural(\alpha(\omega) d(px))$ (where α and p are as in (2)) defines a linear isomorphism $\Omega^{odd} A \xrightarrow{\alpha'} \Omega^1 TA_\natural$.

(4) The isomorphism $\alpha \oplus \alpha' : \Omega A \to X(TA)$ described in (2) and (3) identifies the natural filtration $F^n \Omega A = b\Omega^{n+1} A \oplus \Omega^{n+1} A \oplus \Omega^{n+2} A \oplus \ldots$ on $\Omega(A)$ with the JA-adic filtration

$$F^{2n+1} X(TA) = (JA)^{n+1} \oplus \natural((JA)^{n+1} d(TA) + (JA)^n d(JA))$$

$$F^{2n} X(TA) = ((JA)^{n+1} + [(JA)^n, TA]) \oplus \natural((JA)^n d(TA))$$

on $X(TA)$, see 2.25, where JA is the ideal in TA given as the kernel of the natural homomorphism $TA \to A$.

Remark 2.28. The full algebra $(\Omega A, \circ)$ with Fedosov product is isomorphic to the algebra $QA = A * A$, B.3. The isomorphism sends $x + dx$ to $\iota(x)$ and $x - dx$ to $\bar{\iota}(x)$. Thus, in particular, the algebra TA is isomorphic to the even part, for the natural $\mathbb{Z}/2$-grading, of QA.

As a consequence of the preceding proposition, the X-complex

$$X(TA) : TA \xrightleftharpoons[b]{\natural d} \Omega^1(TA)_\natural$$

is isomorphic to a complex of the form $\Omega^{ev} A \xrightleftharpoons[\beta]{\delta} \Omega^{odd} A$ where the maps δ and β have to be determined.

This is the content of the following key theorem.

Theorem 2.29. *([16]) Let δ, β be as above and $\Omega A \xrightarrow{\kappa} \Omega A$ the Karoubi operator. Then*

(a)
$$\beta(\omega) = b\omega - (1+\kappa)d\omega, \qquad \omega \in \Omega^{odd} A$$

$$\delta(\omega) = \sum_{j=0}^{2n+1} \kappa^j d\omega - \sum_{j=0}^{n-1} \kappa^{2j} b\omega, \ \omega \in \Omega^{2n} A$$

(b) Let P be the spectral projection to 1 for κ (see 2.13). P commutes with b, B and β, δ and
$$P\beta = P(b - \tfrac{2}{N}B)$$
$$P\delta = P(B - b\tfrac{N}{2})$$

(b) follows immediately from (a) by using the equality $P\kappa = P$ and $B = NPd$.

The complex $(1-P)\widehat{\Omega}A$ is contractible, both with the boundary operator $B - b$ and with β and δ. The same holds for the complexes in the associated filtration. For a precise statement we refer to [16], §6.

We still denote by P the projection operator corresponding to P under the isomorphism $\Omega A \cong X(TA)$. Let c be the scaling operator on ΩA which is multiplication by c_q on Ω^q where $c_{2n} = c_{2n+1} = (-1)^n n!$. It follows from 2.29 that $c : PX(TA) \to P\Omega A$ is an isomorphism identifying the differential $\natural d$ and $\dot b$ on $PX(TA)$ with the differential $B - b$ on $P\Omega A$. It obviously is compatible with the filtration F^n on both sides described in 2.27(4).

This proves the following theorem

Theorem 2.30. *Let A be an algebra*

(a) For each n we have
$$HC_n(A) = H_i\Big(X(TA) \,/\, F^n X(TA)\Big)$$
where $H_i, i = 0, 1$ is the homology of the $\mathbb{Z}/2$-graded complex ("supercomplex") $X(TA)/F^n X(TA)$ and $i \equiv n \bmod 2$.

(b) Let $\widehat{\Omega}A = \varprojlim_n \Omega A/F^n \Omega A$ and $\widehat{X}(TA) = \varprojlim_n X(TA)/F^n X(TA)$ be the completions. Then
$$HP_i A = H_i(\widehat{\Omega}A) \cong H_i(\widehat{X}(TA))$$
where $i = 0, 1$ and the differentials on $\widehat{\Omega}A$ and $\widehat{X}(TA)$ are given by $B - b$ and by $(\natural d, \dot b)$, respectively.

Corollary 2.31. *For each n we have*
$$HC_n(A) = H_{i+1}(F^n X(TA)) \qquad HC^n(A) = H_{i+1}((F^n X(TA))')$$

where $H_i, i = 0, 1$ is the homology of the $\mathbb{Z}/2$-graded complex $F^n X(TA)$ and its dual $(F^n X(TA))'$ while $i \equiv n \bmod 2$. In particular, any strong trace τ on $(JA)^n$ (i.e. any functional vanishing on the set of commutators $[(JA)^n, TA]$) defines an element in $HC^{2n-1}A$. This element is represented by the cyclic cocycle f_{2n-1} given by

$$f_{2n-1}(x_0, x_1, \ldots x_{2n-1}) = \tau(\omega(x_0, x_1)\omega(x_2, x_3) \ldots \omega(x_{2n-2}, x_{2n-1}))$$
$$-\tau(\omega(x_{2n-1}, x_0)\omega(x_1, x_2) \ldots \omega(x_{2n-3}, x_{2n-2}))$$

where $\omega(x, y) = p(xy) - p(x)p(y)$ for the canonical linear inclusion $p : A \to TA$.

Proof. The first assertion follows from 2.30 and the exact sequence of complexes $0 \to F^n X(TA) \to X(TA) \to X(TA)/(F^n X(TA)) \to 0$ together with the fact that the homology of $X(TA)$ is trivial.
The second assertion is an easy consequence of the fact that $(JA)^n$ is exactly the even part of $F^{2n-1}X(TA)$.

The preceding corollary is a translation of 2.12. There is a $\mathbb{Z}/2$-graded version using supertraces. A supertrace on a $\mathbb{Z}/2$-graded algebra with grading automorphism $x \to \bar{x}$ is by definition a linear functional f such that $f(xy) = (-1)^{\deg x \deg y} f(yx)$. f is called odd if $f(\bar{x}) = -f(x)$.

Proposition 2.32. ([8]) *Let qA be the canonical ideal in the free product $QA = A * A$ defined in B.3 with its natural $\mathbb{Z}/2$-grading. Any odd supertrace τ on $(qA)^{2n+1}$ defines an element of HC^{2n} represented by the cyclic cocycle*

$$f(x_0, x_1, \ldots, x_{2n}) = \tau(q(x_0)q(x_1) \ldots q(x_{2n}))$$

Here, $q(x)$ is defined for $x \in A$ by $q(x) = \iota(x) - \bar{\iota}(x)$ as in B.3.

Theorem 2.33. *Let $0 \to J \to T \to A \to 0$ be any extension of A where T is quasi-free. Then the complex $X(T)$ is homotopy equivalent to $X(TA)$ by a chain map respecting the filtrations induced by J and JA as in 2.25. Therefore $HC_n(A)$ and $HP_{0/1}A$ can be computed from $X(T)$. Explicitly*

$$HC_n(A) = H_i\Big(X(T)/F^n X(T)\Big) \quad \text{for } i \equiv n \bmod 2$$
$$HP_i A = H_i(\widehat{X}(T)) \quad \text{for } i = 0, 1$$

Proof. This follows from the fact that the X-complex is homotopy invariant for quasi-free algebras (see 2.43) and the fact that T and TA are homotopy equivalent by a homomorphism that respects the filtrations of T and TA by the powers of the ideals J and JA.

Remark 2.34. (a) The operator $S : HC_{n+2} \to HC_n$ is induced by the natural map $\Omega/F^{n+2}\Omega \to \Omega/F^n\Omega$ or by the map $X/F^{n+2}X \to X/F^n X$.

(b) Also the Hochschild homology $HH_n(A)$ can be read off from the filtered complex $X(T)$, if T is any quasi-free extension of A. In fact, $HH_n(A)$ is the homology of the $\mathbb{Z}/2$-graded complex $F^{n-1}XT/F^nXT$ in dimension $i \in \{0,1\}$ with $i \equiv n \bmod 2$.

Corollary 2.35. *(Goodwillie's theorem [20]). Let I be an ideal in the algebra A that is nilpotent, i.e., $I^k = 0$ for some k. Then the quotient map $A \to A/I$ induces an invertible element in $HP_0(A, A/I)$. In particular $HP_*(A) \cong HP_*(A/I)$.*

Proof. We can choose TA as a quasi-free extension for both A and A/I. Since I is nilpotent, the resulting filtrations on $X(TA)$ are continuous with respect to each other.

Remark 2.36. The fact that cyclic and periodic cyclic homology can be reduced to the homology of the X-complex applied to a quasi-free algebra, means that the properties of the theory can be checked by computations that only involve the cyclic homology of a quasi-free algebra *in dimensions 0 and 1*! This fact is essential for establishing some of the most important properties of the theory.

2.6 Cyclic homology as non-commutative de Rham theory

For smooth commutative algebras the cyclic homology/cohomology can be determined by computing first the Hochschild homology and then using Connes's SBI-sequence. This applies in essentially the same way to the coordinate ring of a smooth variety, discussed here, or to the Fréchet algebra of smooth functions on a compact $\mathcal{C}^\infty$-manifold, discussed in 3.2.

Definition 2.37. *A commutative algebra A is smooth if any homomorphism $\alpha : A \to B/N$ where B is a commutative algebra and N an ideal in B with $N^2 = 0$, can be lifted to a homomorphism $\widehat{\alpha} : A \to B$ such that $\pi \circ \widehat{\alpha} = \alpha$ for the quotient map $\pi : B \to B/N$.*

There are various other characterizations of smoothness. Note that, in particular, the coordinate ring of a smooth algebraic variety is smooth.

For a unital commutative algebra we define $\Omega^*_{comm} A$ as the largest graded commutative quotient of ΩA, i.e., as ΩA divided by the ideal generated by all differences $xdy - d(y)x$ and $dxdy + dydx$—with the additional relation $d(1) = 0$. It also can be defined as the exterior algebra over A on the A-bimodule $\Omega^1 A$, with additional relation $d(1) = 0$.

The Hochschild homology of a smooth algebra is determined by the following theorem.

Theorem 2.38. *(Hochschild-Kostant-Rosenberg) If A is a smooth unital commutative algebra then the natural map $\Omega^*_{comm} A \to HH_*(A)$ is an isomorphism (of graded algebras).*

The de Rham cohomology $H^{dR}A$ of A is by definition the homology of the complex
$$\cdots \to \Omega^k_{\text{comm}}A \xrightarrow{d} \Omega^{k+1}_{\text{comm}}A \to \cdots$$
For $* = 0, 1$ we write $H_0^{dR}A$ for the direct sum of all de Rham homology groups in even dimensions and $H_1^{dR}A$ for the sum of all de Rham homology groups in odd dimensions.

Theorem 2.39. *Let A be a unital smooth algebra of homological dimension n. Then*

(a) $HP_*A \cong H^{dR}_*A$
(a) *If $k > n$, then $HC_kA \cong H^{dR}_*A$ for $* \equiv k \bmod 2$.*

Proof. The quotient map $\Omega^n A \to \Omega^n_{\text{comm}}A$ induces a natural map from the (B, b)-bicomplex for A to the bicomplex

$$\begin{array}{ccccc} \downarrow 0 & & \downarrow 0 & & \downarrow 0 \\ \Omega^2_{\text{comm}}A & \xleftarrow{d} & \Omega^1_{\text{comm}}A & \xleftarrow{d} & \Omega^0_{\text{comm}}A \\ \downarrow 0 & & \downarrow 0 & & \\ \Omega^1_{\text{comm}}A & \xleftarrow{d} & \Omega^0_{\text{comm}}A & & \\ \downarrow 0 & & & & \\ \Omega^0_{\text{comm}}A & & & & \end{array}$$

By the Hochschild-Kostant-Rosenberg theorem this map is a quasi-isomorphism on the columns. Thus the theorem follows for instance from the spectral sequence comparison theorem (or also from Connes's SBI-sequence).

Remark 2.40. If A is the coordinate ring for a not necessarily smooth variety V then HP_*A is isomorphic to the infinitesimal cohomology of V in the sense of Grothendieck, which is in this case the correct replacement for de Rham theory.

One can see this is as follows. The infinitesimal cohomology is defined by writing A as a quotient of a smooth algebra S, completing S to $\hat{S}$ and then to take the de Rham cohomology of $\hat{S}$. But now $\hat{S}$ is again a smooth (topological) algebra so that the de Rham cohomology coincides by the argument in 2.39 with the periodic cyclic homology $HP_*(\hat{S})$. Finally, by Goodwillie's theorem $HP_*(\hat{S}) = HP_*(A)$.

All results in this section are of course also valid in cohomology in place of homology.

Remark 2.41. Karoubi has defined de Rham homology for an arbitrary algebra A as the homology of the complex given by the A-bimodules $\Omega^n(A)/[\Omega^*(A), \Omega^*(A)]$, where the quotients are by graded commutators, and has shown that it is isomorphic to the image of $HC_{n+2}(A)$ in $HC_n(A)$ under the operator S, see [27].

There is an even more significant and important analogy between cyclic homology and Grothendieck's notion of infinitesimal homology which describes the topology of non-smooth varieties in algebraic geometry. In this analogy, quasi-free algebras play the role of smooth varieties in the non-commutative category and the X-complex corresponds to the de Rham complex.

In fact, in algebraic geometry, as mentioned above the infinitesimal cohomology is defined by writing the coordinate ring A of the variety as a quotient of the coordinate ring of a smooth variety S and by taking the de Rham cohomology of the completion $\widehat{S}$, cf. 2.40 . It can be shown that this does not depend on the choice of the embedding into a smooth variety.

The periodic cyclic homology of an algebra A is obtained by writing A as a quotient of a quasi-free algebra T and taking the X-complex homology of the completion $\widehat{T}$. This does not depend on the choice of the quasi-free extension T, see 2.30.

2.7 Homotopy invariance for cyclic theory

The important basic features of cyclic homology are homotopy invariance (under differentiable homotopies), Morita invariance and excision. These properties hold very generally for the periodic theory HP—but only in a very limited sense for the non-periodic theory HC.

In this section we discuss homotopy invariance ([4, 20]). The first (and main) step consists in showing that derivations act trivially on HP (or on HC after stabilizing by the S-operator). This can be done by guessing the correct homotopy formula on the complex $(\Omega^n(A), B-b)$ computing the cyclic homology $HC_*(A)$, or, in the X-complex picture, by using the fact that the X-complex is naturally homotopy invariant for quasi-free algebras, [16].

Given an algebra B, we consider the algebraic tensor product $B \otimes \mathcal{C}^\infty[0,1]$ where $\mathcal{C}^\infty[0,1]$ stands for the algebra of $\mathbb{C}$-valued $\mathcal{C}^\infty$-functions on $[0,1]$ (i.e., restrictions of functions in $\mathcal{C}^\infty(\mathbb{R})$ to $[0,1]$).

By a differentiable homotopy from an algebra A to an algebra B we mean a homomorphism α from A to the algebra $B \otimes \mathcal{C}^\infty[0,1]$. Given such a family α, evaluation for each $t \in [0,1]$ defines a homomorphism $\alpha_t : A \to B$. We say that two homomorphisms $\beta_0, \beta_1 : A \to B$ are differentiably homotopic if there exists a differentiable homotopy α such that $\alpha_0 = \beta_0$, $\alpha_1 = \beta_1$.

Remark 2.42. To algebraically minded readers used to polynomial homotopies this definition of homotopy may seem unfamiliar (as it uses an algebra such as $\mathcal{C}^\infty[0,1]$ which is not algebraically defined). It is however obvious that the homotopy invariance result for differentiable homotopies that we will describe is much stronger than just invariance under polynomial homotopies. This stronger result is quite important even in simple applications. For instance homotopies given by rotations in matrices are not polynomial but transcendental involving functions such as sine or cosine. On the other hand the property of homotopy invariance under differentiable homotopies is closely related to the purely algebraic fact that derivations act trivially on periodic cyclic homology.

Let $u : A \to B \otimes C^\infty[0,1]$ be a differentiable family of homomorphisms between algebras A and B. For each fixed $t \in [0,1]$ let $u_t : A \to B$ be the homomorphism given by u at the point t, and $\dot{u}_t$ the derivative, with respect to t, at the point t. This is a linear map $A \to B$ and behaves like a derivation with respect to u_t, i.e.,

$$\dot{u}_t(xy) = u_t(x)\dot{u}_t(y) + \dot{u}_t(x)u_t(y), \quad x, y \in A$$

The Lie derivative
$$L_t = L(u_t, \dot{u}_t) : X(A) \to X(B)$$
is defined by
$$L_t(x) = \dot{u}_t(x), \ L_t(\natural(xdy)) = \natural(\dot{u}_t(x)d(u_t y) + u_t(x)d(\dot{u}_t y)),$$

$x, y \in A$. Then L_t is a morphism of supercomplexes, see [16], beginning of §7.

Proposition 2.43. *Let $u : A \to B \otimes C^\infty[0,1]$ be as above and assume that A is quasi-free. Then, for each $t \in [0,1]$, there is a map of supercomplexes of odd degree $h_t : X(A) \to X(B)$ such that $L_t = \partial h_t + h_t \partial$ where ∂ is the boundary operator in the X-complex.*

Proof. Let φ be the fundamental map $A \to \Omega^2 A$ for the quasi-free algebra A, see (9). Put

$$h_t(x) = \natural i \varphi(x)$$
$$h_t(\natural(xdy)) = i(xdy + b(x\varphi(y)))$$

$x, y \in A$, where

$$i(x_0 dx_1 \ldots dx_k) = u_t(x_0)\dot{u}_t(x_1)du_t(x_2)\ldots du_t(x_k).$$

One computes that $L = \partial h + h \partial$, see [16].

By integrating h_t with respect to t, see [16], 8.1, we obtain an operator $H = \int_0^1 h_t dt$ and

Proposition 2.44. *Let $u : A \to B \otimes C^\infty[0,1]$ be as above and A quasi-free. There is an odd map of supercomplexes $H : X(A) \to X(B)$ such that $u_{1*} - u_{0*} = \partial H + H \partial$ for the maps $u_{t*} : X(A) \to X(B)$ induced by u_t, $t = 0, 1$.*

Assume now that A and B are written as quotients of free algebras T and R, respectively:
$$A \cong T/I \quad B \cong R/J$$
and that $u : A \to B \otimes C^\infty[0,1]$ is a homotopy as before. Since T is free, u induces a homomorphism $w : T \to R \otimes C^\infty[0,1]$ which makes the following diagram commutative:

$$\begin{array}{ccc} T & \xrightarrow{w} & R \otimes \mathcal{C}^\infty[0,1] \\ \downarrow & & \downarrow \\ A & \xrightarrow{u} & B \otimes \mathcal{C}^\infty[0,1] \end{array}$$

Necessarily, w maps I to $J \otimes \mathcal{C}^\infty[0,1]$ and thus each w_t maps I to J. Therefore the map induced by w_t on the X-complexes maps the subspace F^n in the filtration of $X(T)$ defined by I, see 2.25, to the corresponding subspace F^n in the filtration of $X(R)$ given by J. The formulas used in the definition of $L(u_t, \dot{u}_t)$ and H show that these maps send F^n to F^{n-1}. As a consequence, we get

Theorem 2.45. *Let $u : A \to B \otimes \mathcal{C}^\infty[0,1]$ be a differentiable homotopy and $u_t : A \to B$ the corresponding family of homomorphisms.*

(a) The element $HP_0(u_t) \in HP_0(A,B)$ defined by u_t does not depend on t. Thus, in particular, the maps induced by u_t on HP_* and on HP^* are constant.

(b) Let $(u_t)_*$ and $(u_t)^*$ be the maps induced by u_t on HC_n and HC^n and S the periodicity operator on cyclic homology and cohomology. Then $(u_t)_* \circ S : HC_{n+2}(A) \to HC_n(B)$ and $S \circ (u_t)^* : HC^n(A) \to HC^{n+2}(B)$ are constant.

2.8 Morita invariance

The periodic case

Theorem 2.46. *([14]) Let E and F be linear subspaces of the algebra A and $A(EF)$, $A(FE)$ the subalgebras of A generated by the spaces of products EF and FE. Then there is an invertible element in $HP_0(A(EF), A(FE))$.*

This theorem can be proved using the following quasi-free extensions of $A(EF)$ and $A(FE)$:

$$0 \to K(EF) \to T(E \otimes F) \xrightarrow{\pi_1} A(EF) \to 0$$

$$0 \to K(FE) \to T(F \otimes E) \xrightarrow{\pi_2} A(FE) \to 0$$

Here, π_1 and π_2 are the homomorphisms induced by the linear maps $E \otimes F \ni x \otimes y \mapsto xy \in A(EF)$, $F \otimes E \ni y \otimes x \mapsto yx \in A(FE)$ and $K(EF), K(FE)$ are the kernels of π_1, π_2.

The invertible element in $HP_0(A(EF), A(FE))$ is induced by the linear isomorphism

$$\lambda : X(T(E \otimes F)) \to X(T(F \otimes E))$$

which is simply cyclic permutation on tensor products

$$\lambda(x_1 \otimes y_1 \otimes x_2 \otimes y_2 \otimes \ldots \otimes x_k \otimes y_k)$$
$$= y_k \otimes x_1 \otimes y_1 \otimes x_2 \otimes \ldots \otimes x_k$$

$$\lambda(\natural(x_1 \otimes y_1 \otimes \ldots \otimes x_{k-1} \otimes y_{k-1}d(x_k \otimes y_k)))$$
$$= \natural(y_k \otimes x_1 \otimes y_1 \otimes \ldots \otimes x_{k-1}d(y_{k-1} \otimes x_k))$$

$$x_1, \ldots, x_k \in E;\ y_1, \ldots, y_k \in F$$

It is trivially checked that λ commutes with the boundary operators $\natural d$ and $\dot{b}$ of the X-complex, using the definition of these operators.

One can also show that $\lambda : X(T(E \otimes F)) \to X(T(F \otimes E))$ is "continuous" with respect to the natural filtration, see 2.25, i.e., that for each n there is k such that λ maps F^k to F^n, see [14].

A Morita context
$$\begin{pmatrix} A & X \\ Y & B \end{pmatrix}$$
for two algebras A and B is by definition an algebra with a splitting into four linear subspaces such that when the elements are written as 2×2 matrices the multiplication is consistent with matrix multiplication.

Corollary 2.47. *Let A and B be algebras which are related by a Morita context as above and assume that $A = XY$, $B = YX$. Then there is an invertible element in $HP_0(A, B)$.*

The non-periodic case

In the non-periodic theory, Morita invariance only holds for unital or at least H-unital algebras [4, 54].

Theorem 2.48. *Let E and F be linear subspaces of the algebra A and $A(EF)$, $A(FE)$ the subalgebras of A generated by the spaces of products EF and FE. Assume that $A(EF), A(FE)$ are both H-unital (see 2.9). Then there is an invertible element in the bivariant theory $HC^0(A(EF), A(FE))$ (see 2.19). This element induces isomorphisms*
$$HC^n(A(EF)) \cong HC^n(A(FE))$$
for all n.

Corollary 2.49. *Let A and B be H-unital algebras. Let E be a $A-B$-bimodule and F be a $B - A$-bimodule such that*
$$E \otimes_B F \cong A \qquad F \otimes_A E \cong B$$
Then there is an invertible element in $HC^0(A, B)$ inducing isomorphisms
$$HC^n(A) \cong HC^n(B)$$
for all n.

2.9 Excision

Along with the SBI-sequence 2.10, excision is the most important tool for the computation of cyclic homology groups. It allows to compute these invariants by using suitable extensions.

The periodic case

Theorem 2.50. *Let $0 \to S \to P \to Q \to 0$ be an extension of algebras and A an algebra. There are two natural six-term exact sequences*

$$\begin{array}{ccccc} HP_0(A,S) & \to & HP_0(A,P) & \to & HP_0(A,Q) \\ \uparrow & & & & \downarrow \\ HP_1(A,Q) & \leftarrow & HP_1(A,P) & \leftarrow & HP_1(A,S) \end{array}$$

and

$$\begin{array}{ccccc} HP_0(S,A) & \leftarrow & HP_0(P,A) & \leftarrow & HP_0(Q,A) \\ \downarrow & & & & \uparrow \\ HP_1(Q,A) & \to & HP_1(P,A) & \to & HP_1(S,A) \end{array}$$

where the horizontal arrows are induced by the maps in the given extension (a description of the vertical arrows will be given below).

The proof of this theorem given in [17] uses first a reduction to the case where the algebra P is quasi-free. In this case, one can replace P and S by their completions $\hat{P}$ and $\hat{S}$, with respect to the powers of the ideal S (more precisely these completions are to be viewed as pro-algebras). Then $\hat{S}$ is H-unital (cf. 2.9) in an appropriate sense so that Wodzicki's excision argument applies. A simplification of the proof found by R.Meyer, [36], avoids this reduction to the H-unital case. A sketch of Meyer's argument can be found below after Theorem 5.12.

We include a brief discussion of the vertical arrows in 2.50 (see [17] and also [30], III 2.2). Given an extension of algebras $0 \to S \to P \to Q \to 0$ let δ, δ' be the connecting maps in the following two six-term exact sequences.

$$\longrightarrow HP_*(P,S) \longrightarrow HP_*(S,S) \xrightarrow{\delta} HP_{*+1}(Q,S) \longrightarrow$$
$$\longrightarrow HP_*(Q,P) \longrightarrow HP_*(Q,Q) \xrightarrow{\delta'} HP_{*+1}(Q,S) \longrightarrow$$

Let 1_S and 1_Q denote the class of id in $HP_0(S,S)$ and $HP_0(Q,Q)$ respectively.

Proposition 2.51. *One has $\delta(1_S) = -\delta'(1_Q)$.*

Theorem 2.52. *Let $E: 0 \to S \to P \to Q \to 0$ be an extension of algebras. We write $ch(E)$ for the element $\delta'(1_Q) = -\delta(1_S) \in HP_1(Q,S)$. The connecting map in the first exact sequence in Theorem 5.3 acts on $HP_j(A,Q)$ by multiplication by $(-1)^j ch E$ on the right, while the connecting map in the second exact sequence in Theorem 5.3 is given by multiplication by $-ch(E)$ on the left.*

The non-periodic case

Let I be a (non-unital) algebra and $(\Omega^n(I), b)$ and let $(D_n^\Omega(I), B'-b)$ be the standard complexes computing Hochschild and cyclic homology of I, respectively, see 2.14. If $0 \to I \to A \to B \to 0$ with I as ideal, we let $(\Omega^n(I:A), b)$ and $(D_n^\Omega(I:A), B'-b)$ denote the kernels of the maps $\Omega^n(A) \to \Omega^n(B)$ and $D_n^\Omega(A) \to D_n^\Omega(B)$, respectively, so that we have exact sequences of complexes:

$$0 \to \Omega^n(I:A) \to \Omega^n(A) \to \Omega^n(B) \to 0$$
$$0 \to D_n^\Omega(I:A) \to D_n^\Omega(A) \to D_n^\Omega(B) \to 0$$

We say that I is *excisive* in Hochschild homology (respectively in cyclic homology), if the inclusion $\Omega^n(I) \to \Omega^n(I:A)$ (respectively $D_n^\Omega(I) \to D_n^\Omega(I:A)$) is a homotopy equivalence of complexes. If this is the case then obviously the exact sequences of complexes above induce long exact sequences in Hochschild (respectively cyclic) homology.

Theorem 2.53. *([53, 54]) Let I be a (non-unital) algebra. The following conditions are equivalent:*

(a) I is H-unital (see 2.9 for the definition).
(b) I is excisive in Hochschild homology, i.e., any extension of the form $0 \to I \to A \to B \to 0$ induces a long exact sequence in Hochschild homology

$$\ldots HH_n(I) \to HH_n(A) \to HH_n(B) \to HH_{n-1}(I)$$
$$\to HH_{n-1}(A) \to HH_{n-1}(B) \to \ldots$$

(c) I is excisive in cyclic homology, i.e., any extension of the form $0 \to I \to A \to B \to 0$ induces a long exact sequence in cyclic homology

$$\ldots HC_n(I) \to HC_n(A) \to HC_n(B) \to HC_{n-1}(I)$$
$$\to HC_{n-1}(A) \to HC_{n-1}(B) \to \ldots$$

A simple proof can be found in [22].

2.10 The Chern character for K-theory classes given by idempotents and invertibles

With homotopy classes of idempotents and invertibles in a matrix algebra over A one can associate periodic cyclic homology classes as follows.

Let A be unital, e an idempotent in a matrix algebra $M_n(A)$ and $[e]$ its homotopy class. e may be thought of as a homomorphism $\mathbb{C} \to M_n(A)$. It induces a map $HP_0\mathbb{C} \to HP_0(M_n(A)) \cong HP_0 A$. We denote by $ch([e])$ the image of the generator of $HP_*\mathbb{C} \cong \mathbb{C}$ (see 2.24) under this map.

Similarly, an invertible element u in $M_n(A)$ may be thought of as a map from the algebra $\mathbb{C}[z, z^{-1}]$ to $M_n(A)$ mapping z to u. This induces a map $HP_1\mathbb{C}[z, z^{-1}] \to HP_1(M_n(A)) \cong HP_1 A$. We denote by $ch([u])$ the image in $HP_1 A$ of the generator of $HP_1\mathbb{C}[z, z^{-1}] \cong \mathbb{C}$ (see 2.24) under this map.

Remark 2.54. It is very easily seen that the classes $ch([e])$ and $ch([u])$ are additive for the natural K-theory addition. Therefore one may view these classes as defining homomorphisms $ch : K_*^{alg} A \to HP_*A$, $* = 0, 1$ from the algebraic K-theory to HP_* (which in fact can be extended to the higher-dimensional algebraic K-theory in a rather straightforward way). This homomorphism even passes to the quotient of algebraic K-theory by homotopy. Since there is no good definition of topological K-theory for arbitrary m-algebras based only on idempotents and invertibles in matrix algebras, it is not clear, a priori, how to interpret this as a map from topological K-theory though.

Such a map from topological K-theory is obtained as a special second variable case of the bivariant character described in Chap. 3.

To obtain explicit formulas for the cycles $ch([e])$ and $ch([u])$ in the periodic cyclic complex it is very convenient to use the description of $HP_*(A)$ as the homology of the completed X-complex $\widehat{X}(TA)$ and its homotopy equivalence with the (B, b)-bicomplex $\widehat{\Omega}A$.

The idempotent e admits a canonical lifting to an idempotent $\hat{e}$ in each nilpotent extension, and therefore in the projective limit $M_n(\widehat{T}A)$ of nilpotent extensions of $M_n(A)$. This is easily seen by putting $f = 2e-1$ in the unitization and defining a lift $\hat{f}$ for f by the formula $\hat{f} = \sigma(f)(\sigma(f)^2)^{-1/2}$, given by a power series in the nilpotent variable $1 - \sigma(f)^2$ (σ denotes the canonical linear lift $M_n(A) \to M_n(TA)$). Under the isomorphism $M_n(\widehat{T}A) \cong M_n(\widehat{\Omega}^{ev}A)$ with Fedosov product, 2.27, $\hat{e}$ is given by the following formula

$$\hat{e} = e + \sum_{n \geq 1} \frac{(2n)!}{(n!)^2}\left(e - \frac{1}{2}\right)(de)^{2n}$$

where de is to be considered as the matrix (de_{ij}) if $e = (e_{ij})$. Now, $\text{Tr}(\hat{e})$ is an even cycle in the complex $\widehat{X}(TA) = X(\widehat{T}A)$, representing $ch([e])$ (since the generator of $HP_0(\mathbb{C})$ is represented by 1 in $X(\mathbb{C})$). Under the identification of $\widehat{X}(TA)$ with $\widehat{\Omega}A$, this becomes

$$ch([e]) = \text{Tr}(e) + \sum_{n \geq 1}(-1)^n \frac{(2n)!}{n!}\text{Tr}\left(\left(e - \frac{1}{2}\right)(de)^{2n}\right)$$

as an element of the complex $\widehat{\Omega}A$ with boundary operator $B - b$.

Let further u be an invertible element in the unitization $M_n(A)\widetilde{\ }$ and consider the reduced X-complex $\overline{X}(M_n(A)\widetilde{\ })$ where we introduce the relation $d(1) = 0$ for the adjoined unit 1. By a similar argument as before, the element $\natural(u^{-1}du)$ can be lifted to a cycle $\natural(\hat{u}^{-1}d\hat{u})$ in $\overline{X}(M_n(\widehat{T}A)\widetilde{\ })$. The trace on M_n induces a map $\text{Tr} : \overline{X}(M_n(\widehat{T}A)\widetilde{\ }) \to \overline{X}(\widehat{T}A\widetilde{\ })$ and $\text{Tr}(\natural(\hat{u}^{-1}d\hat{u}))$ will represent $ch([u])$ (the generator of $HP_1(\mathbb{C}[z, z^{-1}])$ is represented by $\natural(z^{-1}dz)$ in $X(\mathbb{C}[z, z^{-1}])$). Then $ch([u])$ corresponds to

$$ch([u]) = \sum_{n \geq 0} n!\, \text{Tr}((u^{-1} - 1)d(u - 1)\left(d(u^{-1} - 1)d(u - 1)\right)^n)$$

These formulas for ch([e]) and ch([u]) can also be obtained by straightforward computation in the $B - b$ - bicomplex and have been noted by various authors. The classes ch([e]) and ch([u]) have first been constructed in [4,27]. The conceptual interpretation given here is taken from [16].

3 Cyclic Theory for locally convex algebras

By a locally convex algebra we mean an algebra $\mathcal{A}$ over $\mathbb{C}$ equipped with a locally convex topology for which the multiplication is (jointly) continuous. This means that the topology on $\mathcal{A}$ is defined by a family $\mathcal{P}$ of seminorms such that for each $p \in \mathcal{P}$ there is $q \in \mathcal{P}$ such that

$$p(xy) \leq q(x)q(y) \qquad x,y \in \mathcal{A}$$

All our locally convex algebras will be assumed to be complete.

Basically all definitions, constructions and results in the previous chapter on cyclic theory carry over to the category of locally convex algebras. The only point where additional care and work really is necessary is the proof of excision sketched in 2.9. This is due to the fact that the proof in [17] uses the canonical free extension of $\mathcal{A}$ given by the tensor algebra over $\mathcal{A}$. This standard free extension does not exist for an arbitrary locally convex algebra. The difficulty can however be overcome, cf. [10].

The arguments work more smoothly and in fact can be carried over directly from the algebraic case for a somewhat more restricted class of locally convex algebras, which we call m-algebras for brevity. An m-algebra is a locally convex algebra $\mathcal{A}$ with topology given by a family $\mathcal{P}$ of seminorms such that for each $p \in \mathcal{P}$

$$p(xy) \leq p(x)p(y) \qquad x,y \in \mathcal{A}$$

(i.e. all p are submultiplicative). For an m-algebra $\mathcal{A}$, the tensor algebra $T\mathcal{A}$ can be equipped with a completely canonical m-algebra topology which makes it a free extension of $\mathcal{A}$.

Since the category of m-algebras also is the category on which a bivariant K-theory can be developed smoothly (one main ingredient again is the free extension given by the tensor algebra), see Chap. 4, we will restrict here to this class. Note that the class is quite large and actually contains the algebras arising typically in applications from differential geometry.

The necessary (easy) background on locally convex algebras and m-algebras as well as some standard examples of m-algebras, that will be used in the following two chapters, will be treated in the appendix.

3.1 General modifications

To obtain the appropriate definitions of cyclic homology and cohomology in the category of m-algebras, in general of course all morphisms should be continuous, duals should be topological duals. Note also that in the definition of the homology of a complex we divide by the image of the boundary operator and **not** by its closure (in many cases the image is automatically closed).

We now list some more specific points

- All tensor products, e.g. in the cyclic bicomplex or the Hochschild complex but also elsewhere will be completed projective tensor products (see Appendix A). The dual of such a tensor product then consists exactly of the continuous multilinear functionals.
- If $\mathcal{A}$ is an m-algebra, then the algebra of differential forms $\Omega\mathcal{A}$ (see 2.3) is defined using the completed projective tensor product

$$\Omega^n \mathcal{A} = \widetilde{\mathcal{A}} \hat{\otimes} \mathcal{A}^{\hat{\otimes} n} \cong \mathcal{A}^{\hat{\otimes}(n+1)} \oplus \mathcal{A}^{\hat{\otimes} n}$$

and

$$\Omega\mathcal{A} = \bigoplus \Omega^n \mathcal{A} \qquad \hat{\Omega}\mathcal{A} = \prod \Omega^n \mathcal{A}$$

the direct sum and product being taken in the category of locally convex spaces.

The operators on $\Omega\mathcal{A}$ considered in 2.3 are of course continuous and extend to the completed tensor products.

- The n-th power $\mathcal{A}^n$ of an m-algebra is defined to be the image (in $\mathcal{A}$) of the n-fold tensor power $\mathcal{A}^{\hat{\otimes} n}$ under the multiplication map.
- The characterizations of quasi-free algebras in 2.22 carry over directly to the category of m-algebras. In condition (1) of 2.22 N^k has to be interpreted as the image of the n-fold projective tensor power of N under the multiplication map and the nilpotent extension has to be assumed to admit a continuous linear splitting.

The tensor algebra $T\mathcal{A}$ for an m-algebra $\mathcal{A}$ is the natural m-algebra completion of the algebraic tensor algebra over $\mathcal{A}$ described in Appendix A.2. It is quasi-free as an m-algebra.

In the definition of cyclic homology using the X-complex of a quasi-free resolution and in particular in the definition of the filtration of the X-complex in 2.25 it is important to note the following property of ideals in quasi-free m-algebras:

If $\mathcal{I}$ is the ideal in an extension

$$0 \longrightarrow \mathcal{I} \longrightarrow \mathcal{A} \longrightarrow \mathcal{A}/\mathcal{I} \longrightarrow 0$$

of m-algebras with a continuous splitting and $\mathcal{A}$ is quasi-free, then also $\mathcal{I}^n$ is an ideal which is a direct summand in $\mathcal{A}$ as a topological vector space for each n. For instance, $(J\mathcal{A})^n$ is a topological direct summand in $T\mathcal{A}$ for each n.

- In the definition of differentiable homotopy in 2.7 we use, for m-algebras, the completed projective tensor product $\mathcal{C}^\infty[0,1]\hat{\otimes}\mathcal{B}$. Note that $\mathcal{C}^\infty[0,1]\hat{\otimes}\mathcal{B}$ can be identified with the algebra of B-valued $\mathcal{C}^\infty$-functions on $[0,1]$.

- The results on Morita invariance in 2.8 and 2.8 are also valid for m-algebras replacing $A(EF)$, $A(FE)$, XY and YX by the closed subalgebras generated by the respective products.
- All extensions of m-algebras are assumed to admit a continuous linear splitting. For such extensions the excision theorems in 2.9 and 2.9 are valid, see [12] and [54].
- The algebra $q\mathcal{A}$ used in 2.32 is in the case of an m-algebra $\mathcal{A}$ the algebra constructed in B.3.

With these provisions all the results in Chap. 1 carry over to m-algebras. This holds in particular for the SBI-sequence, for homotopy invariance, Morita invariance and for excision.

3.2 De Rham theory for differentiable manifolds

Let M be a differentiable manifold. To treat the case of the Fréchet algebra $\mathcal{C}^\infty M$ of course we use the cyclic homology for locally convex algebras. There is the following analogue of the Hochschild-Kostant-Rosenberg theorem, 2.38.

Theorem 3.1. *(Connes) Let $\mathcal{C}^\infty M$ be the algebra of smooth functions on a smooth compact manifold M and let $\mathcal{C}^\infty(\Lambda T^*M)$ be the algebra of differential forms on M. Then the natural map $\mathcal{C}^\infty(\Lambda^k T^*M) \to HH_k(\mathcal{C}^\infty M)$ is an isomorphism for each k.*

Theorem 3.2. *Let M be a smooth compact manifold of dimension n. Then*

(a) $HP_*(\mathcal{C}^\infty M) \cong H^{dR}_*(\mathcal{C}^\infty M)$
(b) If $k > n$, then $HC_k(\mathcal{C}^\infty M) \cong H^{dR}_*(\mathcal{C}^\infty M)$ for $* \equiv k \bmod 2$.

Using 3.1 the proof is completely analogous to the one in the algebraic case sketched in 2.6.

3.3 Cyclic homology for Schatten ideals

The cyclic homology of the m-algebra $\ell^p(H)$, see A.5, can – somewhat unexpectedly – be computed using the excision property of periodic cyclic homology.

Proposition 3.3. *Let $\mathcal{I}$ and $\mathcal{A}$ be m-algebras. We assume that there are continuous homomorphisms $\alpha : \mathcal{I} \to \mathcal{A}$ and $\mu : \mathcal{A}\hat{\otimes}\mathcal{A} \to \mathcal{I}$ with the following properties:*

(a) $\alpha \circ \mu$ is the multiplication on $\mathcal{A}$.
(b) $\mu \circ (\alpha \otimes \alpha)$ is the multiplication on $\mathcal{I}$.

(thus, in particular $\alpha(\mathcal{I})$ is an ideal in $\mathcal{A}$ with $\mathcal{A}^2 \subset \alpha(\mathcal{I})$). Then α induces an invertible element $HP_0(\alpha)$ in $HP_0(\mathcal{I},\mathcal{A})$.

Proof. This follows by applying the excision theorem in HP for m-algebras to the following extension

$$0 \to \mathcal{I}(0,1) \to \mathcal{I}(0,1) + \mathcal{A}\, t \to \mathcal{A} \to 0 \tag{1}$$

Here we deviate from our usual notation and denote by $\mathcal{I}(0,1)$ the algebra of differentiable $\mathcal{I}$-valued functions on $[0,1]$ that vanish at the endpoints, without assuming that their derivatives also vanish. The m-algebra $\mathcal{I}(0,1) + \mathcal{A}\, t$ is defined in the following way. As a locally convex vector space it simply is the direct sum of $\mathcal{I}(0,1)$ and $\mathcal{A}$. The symbol t denotes the identity function on $[0,1]$. The elements of $\mathcal{A}\, t$ are regarded as functions on $[0,1]$ with values in $\mathcal{A}$, which are multiples of t by elements of $\mathcal{A}$. The multiplication on the first summand is the one of $\mathcal{I}(0,1)$. The product of a function f in $\mathcal{I}(0,1)$ by an element $xt \in \mathcal{A}\, t$ is $\mu(\alpha(f) \otimes xt)$ (we extend μ and α canonically to functions). The product of xt and yt in the second factor is defined as $\mu(x \otimes y)(t^2 - t) + \alpha\mu(x \otimes y)t$, where the first summand is in $\mathcal{I}(0,1)$ and the second one in $\mathcal{A}\, t$. The extension constructed that way obviously has a continuous linear splitting and defines an element u in $HP_1(\mathcal{A}, \mathcal{I}(0,1)) \cong HP_0(\mathcal{A}, \mathcal{I})$. This element is inverse to $HP_0(\alpha)$.

We now apply this result to the Schatten ideals $\ell^p = \ell^p(H)$, $p \geq 1$. More generally, consider the case where $\mathcal{I} = \ell^p \hat{\otimes} \mathcal{B}$ and $\mathcal{A} = \ell^q \hat{\otimes} \mathcal{B}$ for an arbitrary m-algebra $\mathcal{B}$ and $p \leq q \leq 2p$. The maps α and μ are obtained from the continuous inclusion $\ell^p \to \ell^q$ and from the multiplication map $\ell^q \hat{\otimes} \ell^q \to \ell^p$.

Corollary 3.4. *For any m-algebra $\mathcal{B}$, $\ell^p \hat{\otimes} \mathcal{B}$ is equivalent in HP_* to $\ell^q \hat{\otimes} \mathcal{B}$, i.e., the map $\ell^p \hat{\otimes} \mathcal{B} \to \ell^q \hat{\otimes} \mathcal{B}$ induced by the inclusion $\ell^p \to \ell^q$ gives an invertible element in $HP_0(\ell^p \hat{\otimes} \mathcal{B}, \ell^q \hat{\otimes} \mathcal{B})$ for $q \geq p$.*

Proposition 3.5. *The inclusion map $\mathbb{C} \to \ell^1(H)$ (which maps $1 \in \mathbb{C}$ to a one-dimensional projector in ℓ^1) and the canonical trace $\mathrm{Tr} : \ell^1(H) \to \mathbb{C}$ define elements in $HP_0(\mathbb{C}, \ell^1(H))$ and $HP_0(\ell^1(H), \mathbb{C})$, respectively, which are inverse to each other. Thus in particular $HP_*(\ell^1(H)) \cong HP_*(\mathbb{C})$.*

Corollary 3.6. *The inclusion map $\mathbb{C} \to \ell^p(H)$ defines an invertible element in $HP_0(\mathbb{C}, \ell^p(H))$ for each $p \geq 1$. Thus in particular $HP_*(\ell^p(H)) \cong HP_*(\mathbb{C})$.*

Remark 3.7. Assume (for simplicity) that p is an odd integer. The generator of $HP^0(\ell^p(H)) = \varinjlim HC^{2n}(\ell^p(H))$ is represented by the cyclic cocycle f in C_λ^{p-1} given by

$$f(x_0, x_1, \ldots, x_{p-1}) = \mathrm{Tr}\,(x_0 x_1 \ldots x_{p-1})$$

It follows from the SBI-sequence that the class of f is not in the image of the S-operator and thus that $p-1$ is the smallest integer k for which the map $HC^k(\ell^p(H)) \to HP^0(\ell^p(H))$ is non-trivial.

3.4 Cyclic cocycles associated with Fredholm modules

The concept of a Fredholm module over an algebra A has been introduced by Atiyah and Kasparov as a framework for K-homology. The bivariant version (over two algebras A and B) is due to Kasparov and is usually called a Kasparov module. It describes an extension of B by an algebra Morita equivalent to A which admits an inverse (i.e, there exists another extension such that its sum with the given extension is equivalent to a trivial extension). The notion of an unbounded Fredholm module or a spectral triple plays a fundamental role in Alain Connes's "dictionary" for noncommutative geometry. We treat in this section only the simplest but typical case of a "bounded" Fredholm module.

Given a Hilbert space H, we denote by $\mathcal{L}(H)$ the algebra of bounded operators on H and by $\ell^p(H)$ the Schatten ideal of p-summable operators in $\mathcal{L}(H)$, A.5.

Definition 3.8. *Let A be an algebra.*

(a) A p-summable odd Fredholm module over A is a representation $\pi : A \to \mathcal{L}(H)$ together with an operator $F \in \mathcal{L}(H)$ such that $F = F^*$, $F^2 = 1$ and such that the commutators $[F, \pi(x)]$, $x \in A$ are all in $\ell^p(H)$.

(b) A p-summable even Fredholm module over A is a representation $\pi : A \to \mathcal{L}(H)$ together with an operator $F \in \mathcal{L}(H)$ such that $F = F^*$, $F^2 = 1$ and $[F, \pi(x)] \in \ell^p(H)$, $x \in A$, and a grading operator γ on H, $\gamma = \gamma^*$, $\gamma^2 = 1$, such that $\gamma F \gamma = -F$ and $\gamma \pi(x) \gamma = \pi(x)$, $x \in A$ (i.e., F has degree 1, while $\pi(A)$ has degree 0 for the $\mathbb{Z}/2$-grading induced by γ).

An odd p-summable Fredholm module determines an extension of the following form

$$0 \longrightarrow \ell^{p/2}(H) \longrightarrow (\ell^{p/2}(H) + \sigma(A)) \longrightarrow A_0 \longrightarrow 0 \qquad (2)$$

where $\sigma(x) = P\pi(x)P$, $P = 1/2(F+1)$ the spectral projection for 1 and $A_0 = A/\mathrm{Ker}(\sigma)$. Note that we can work with $p/2$ rather than p, since $\sigma(xy) - \sigma(x)\sigma(y) = (1/2)P[F,x][F,y]P \in \ell^{p/2}(H)$.

We know from 3.7 that $HP^0(\ell^{2n+1}(H))$ is one-dimensional with generator the image $\dot{g}$ in HP^0 of the cyclic cocycle $g(x_0, \ldots, x_{2n}) = \mathrm{Tr}(x_0 x_1 \ldots x_{2n})$. By excision for HP^*, the image of $\dot{g}$ under the boundary map in the long exact sequence associated to the extension (2) gives an element $ch(A, F)$ in $HP^1(A)$ which is called the character of the Fredholm module (A, F). A more refined version of this argument is the following. By B.2, the extension (2) determines a classifying map $\gamma : JA \to \ell^p(H)$ which maps the p-th power $(JA)^p$ of JA to the p-th power of $\ell^p(H)$, i.e., to $\ell^1(H)$. Composition of γ with the trace on $\ell^1(H)$ gives a trace τ on $(JA)^p$. Now, by 2.31, any trace on $(JA)^p$ determines for each n with $2n+2 \geq p$ cyclic cocycles $f_{2n+1} \in C_\lambda^{2n+1}(A)$ which are related by the S-operator. According to 2.31 the explicit formula for f_{2n+1} is

$$f_{2n+1}(x_0, x_1, \ldots x_{2n+1}) = \mathrm{Tr}(\omega(x_0, x_1)\omega(x_2, x_3) \ldots \omega(x_{2n}, x_{2n+1}))$$
$$- \mathrm{Tr}(\omega(x_{2n+1}, x_0)\omega(x_1, x_2) \ldots \omega(x_{2n-1}, x_{2n}))$$

where $\omega(x,y) = \sigma(xy) - \sigma(x)\sigma(y) = P\pi(xy)P - P\pi x P\pi y P$. A simple computation shows that this formula can be rewritten in terms of commutators $[F,\cdot]$ as follows

$$f_{2n+1}(x_0, x_1, \ldots x_{2n+1}) = \mathrm{Tr}\,(F[F, x_0][F, x_1]\ldots [F, x_{2n+1}])$$

It follows from the proof of excision in HP^* that $ch(A, F)$ is the image of $\frac{(-1)^n}{n!} f_{2n+1}$ under the map $HC^{2n+1} \to HP^1$. There is a simple index theorem based on this cocycle. Recall that a Fredholm operator on a Hilbert space is a bounded operator which is invertible modulo compact operators and that the index of a Fredholm operator L is defined as $\mathrm{ind}\, L = \dim \mathrm{Ker}\, L - \dim \mathrm{Coker}\, L$.

Theorem 3.9. *(Connes) Let A be unital, $u \in A$ invertible, (H, F) a p-summable unital Fredholm module over A and $P = 1/2(F+1)$. Then $u_P = P\pi(u)P$ is a Fredholm operator on the Hilbert space PH. Its index is given by*

$$\mathrm{ind}\, u_P = \langle ch(A, F)|ch(u)\rangle$$
$$= f_{2n+1}(u^{-1}, u, u^{-1}, \ldots, u)$$

With obvious modifications, the theorem is still valid for invertibles in a matrix algebra over A.

An even p-summable Fredholm module determines an extension of the form

$$0 \longrightarrow \ell^p(H) \longrightarrow (\ell^p(H) + \pi(A)) \longrightarrow A \longrightarrow 0$$

with two different splittings $\pi, \bar{\pi} : A \to (\ell^p(H) + \pi(A))$ where $\bar{\pi}(x) = F\pi(x)F$ (we assume here without restriction of generality that π is injective). Using excision for HP we see that $HP^0(\ell^p(H) + \pi(A)) \cong HP^0(\ell^p(H)) \oplus HP^0(A)$. Denote by $ch(A, F)$ the image of 1 under the map

$$\mathbb{C} = HP^0(\ell^p(H)) \xrightarrow{\pi^* - \bar{\pi}^*} HP^0(A)$$

Again there is an explicit formula for a cyclic cocycle representing $ch(A, F)$. In fact, the universality property of the extension

$$0 \longrightarrow qA \longrightarrow QA \longrightarrow A \longrightarrow 0$$

cf. B.3, shows that there is a natural homomorphism $qA \to \ell^p(H)$. Composing again with the supertrace on $\ell^1(H)$ (recall that H is graded) we obtain an odd supertrace on $(qA)^p$ and thus by 2.32 cyclic cocycles. They are determined, for each n with $2n + 1 \geq p$, by the formula

$$f_{2n}(x_0, x_1, \ldots x_{2n}) = \mathrm{Tr}(\gamma q(x_0) q(x_1) \ldots q(x_{2n}))$$

where $q(x) = \pi(x) - \bar{\pi}(x)$ and γ is the grading automorphism on H. One may rewrite this as

$$f_{2n}(x_0, x_1, \ldots x_{2n}) = \mathrm{Tr}\,(\gamma F([F, x_0][F, x_1]\ldots [F, x_{2n}]))$$

The class $ch(A, F)$ is represented by $\frac{(-1)^n n!}{(2n)!} f_{2n}$

Theorem 3.10. *(Connes) Let A be unital, $e \in A$ idempotent and (A, F) a p-summable Fredholm module over A. Then the $F_e = \pi(e)F\pi(e)$ is a Fredholm operator from the Hilbert space $\pi(e)H_+$ to the Hilbert space $\pi(e)H_-$, where $H = H_+ \oplus H_-$ is the eigenspace decomposition of H with respect to the grading operator γ. Its index is given by*

$$\operatorname{ind} F_e = \langle ch(A, F) | ch(e) \rangle$$
$$= f_{2n}(e, e, \ldots, e)$$

In order to promote the character defined for Fredholm modules in this section to a transformation from a K-homology group (with good properties) to (periodic) cyclic cohomology one has to use the first variable case of the bivariant character described in the next chapter.

4 Bivariant K-Theory

A natural category on which topological K-theory has been defined traditionally with good properties, is the one of Banach algebras. It has been studied with particular intensity on the category of C*-algebras. On the category of C*-algebras Kasparov, [28], extended K-theory to a bivariant theory KK in such a way that, for a C*-algebra A, one has $K_*(A) = KK_*(\mathbb{C}, A)$, $* = 0, 1$. A similar bivariant theory on the category of C*-algebras is the E-theory of Connes-Higson, [9]. It is an alternative extension of K-theory to a bivariant theory. In particular it also satisfies $K_*(A) = E_*(\mathbb{C}, A)$, $* = 0, 1$.

Now it is a well-known fact that cyclic homology is degenerate on Banach or C*-algebras. Thus, for instance, if we define cyclic and periodic cyclic homology for a locally convex algebra as in Sect. 2.4, we obtain for the algebra $\mathcal{C}(X)$ of continuous functions on the compact space X

$$HP^0(\mathcal{C}(X)) = \{\text{ bounded complex measures on } X\}$$
$$HP^1(\mathcal{C}(X)) = 0$$

It is possible to define a generalized cyclic homology theory (the local theory discussed in Chap. 4) even in the bivariant setting with good properties on the category of C*-algebras (or Banach algebras) and to define a bivariant transformation from $KK(A, B)$ or from $E(A, B)$ into this theory, see Chap. 4.

However, it is also possible to construct a bivariant topological K-theory on the category of locally convex algebras and then to define a bivariant Chern-Connes character from this theory into the ordinary bivariant cyclic theory $HP_*(A, B)$, as we discussed it in Sect. 2.4. We will describe the construction in this section. As before we work within the category of m-algebras (projective limits of Banach algebras), rather than with general locally convex algebras.

The theory developed in [13], that we present here, is based on extensions of such algebras of arbitrary length. The all-important product will then be induced by the familiar Yoneda product (concatenation of extensions). For technical reasons however it is impossible to work with the extensions directly. Rather we will associate a *classifying map* with each such extension and then consider homotopy classes of such maps.

4.1 Bivariant K-theory for locally convex algebras

Since cyclic theory is homotopy invariant only for differentiable homotopies (called diffotopies below), we have to set up the theory in such a way that it uses only diffotopies in place of general homotopies.

For an m-algebra $\mathcal{A}$ we write $\mathcal{A}[a,b], \mathcal{A}[a,b)$ and $\mathcal{A}(a,b)$ for the m-algebras $\mathcal{A}\hat{\otimes}\mathbb{C}[a,b], \mathcal{A}\hat{\otimes}\mathbb{C}[a,b)$ and $\mathcal{A}\hat{\otimes}\mathbb{C}(a,b)$, see A.1. The elements of these algebras are $\mathcal{A}$-valued functions whose derivatives vanish in 0 and 1.

Definition 4.1. *(see A.1) Two continuous homomorphisms $\alpha, \beta : \mathcal{A} \to \mathcal{B}$ between two m-algebras are called differentiably homotopic or **diffotopic**, if there is a family $\varphi_t : \mathcal{A} \to \mathcal{B}, t \in [0,1]$ of continuous homomorphisms, such that $\varphi_0 = \alpha, \varphi_1 = \beta$ and such that the map $t \mapsto \varphi_t(x)$ is infinitely differentiable for each $x \in \mathcal{A}$.*

According to Appendix A.1, α and β are diffotopic if and only if there exists a continuous homomorphism $\varphi : \mathcal{A} \to \mathbb{C}[0,1]\hat{\otimes}\mathcal{B}$ with the property that $\varphi(x)(0) = \alpha(x), \varphi(x)(1) = \beta(x)$ for each $x \in \mathcal{A}$. This shows that diffotopy is an equivalence relation.

Definition 4.2. *Let $\mathcal{A}$ and $\mathcal{B}$ be m-algebras. For any homomorphism $\varphi : \mathcal{A} \to \mathcal{B}$ of m-algebras, we denote by $\langle\varphi\rangle$ the equivalence class of φ with respect to diffotopy and we set*

$$\langle \mathcal{A}, \mathcal{B} \rangle = \{\langle\varphi\rangle | \varphi \text{ is a continuous homomorphism } \mathcal{A} \to \mathcal{B}\}$$

Let $\mathcal{K}$ be the algebra of smooth compact operators defined in A.4. For two continuous homomorphisms $\alpha, \beta : \mathcal{A} \to \mathcal{K}\hat{\otimes}\mathcal{B}$ we define the direct sum $\alpha \oplus \beta$ as

$$\alpha \oplus \beta = \begin{pmatrix} \alpha & 0 \\ 0 & \beta \end{pmatrix} : \mathcal{A} \longrightarrow M_2(\mathcal{K}\hat{\otimes}\mathcal{B}) \cong \mathcal{K}\hat{\otimes}\mathcal{B}$$

With the addition defined by $\langle\alpha\rangle + \langle\beta\rangle = \langle\alpha\oplus\beta\rangle$ the set $\langle\mathcal{A}, \mathcal{K}\hat{\otimes}\mathcal{B}\rangle$ of diffotopy classes of homomorphisms from $\mathcal{A}$ to $\mathcal{K}\hat{\otimes}\mathcal{B}$ is an abelian semigroup with 0-element $\langle 0 \rangle$ (this follows from A.1).

Recall from B.2 that we denote by $T\mathcal{A}$ the tensor algebra over $\mathcal{A}$ and by $J\mathcal{A}$ the kernel of the natural homomorphism: $T\mathcal{A} \to \mathcal{A}$. Recall also that any extension

$$0 \to \mathcal{I} \to \mathcal{E} \to \mathcal{A} \to 0$$

admitting a continuous linear splitting can be compared with the free extension $0 \to J\mathcal{A} \to T\mathcal{A} \to \mathcal{A} \to 0$ and thus yields a classifying map

$\gamma : J\mathcal{A} \to \mathcal{I}$ which is unique up to homotopy, B.2. For an extension admitting a homomorphism splitting the classifying map is homotopic to 0.

An extension of $\mathcal{A}$ of length n is an exact complex of the form

$$0 \to \mathcal{I} \to \mathcal{E}_1 \to \mathcal{E}_2 \ldots \to \mathcal{E}_n \to \mathcal{A} \to 0$$

where the arrows or boundary maps (which we denote by φ) are continuous homomorphisms between m-algebras. We call such an extension linearly split if there is a continuous linear map s of degree -1 on this complex such that $s\varphi + \varphi s =$ id. This is the case if and only if the given extension is a Yoneda product (concatenation) of n linearly split extensions of length 1: $\mathcal{I} \to \mathcal{E}_1 \to \operatorname{Im}\varphi_1$, $\operatorname{Ker}\varphi_2 \to \mathcal{E}_2 \to \operatorname{Im}\varphi_2$, ...

$J\mathcal{A}$ is, for each m-algebra $\mathcal{A}$, again an m-algebra. By iteration we can therefore form $J^2\mathcal{A} = J(J\mathcal{A}), \ldots, J^n\mathcal{A} = J(J^{n-1}\mathcal{A})$.

Proposition 4.3. *(see B.2) For any linearly split extension*

$$0 \to \mathcal{I} \to \mathcal{E}_1 \to \mathcal{E}_2 \ldots \to \mathcal{E}_n \to \mathcal{A} \to 0$$

of $\mathcal{A}$ of length n there is a classifying map $\gamma : J^n\mathcal{A} \to \mathcal{I}$ which is unique up to homotopy.

Proposition 4.4. *(a) J (and J^n) is a functor, i.e. each continuous homomorphism $\mathcal{A} \to \mathcal{B}$ between m-algebras induces a continuous homomorphism $J\mathcal{A} \to J\mathcal{B}$.*
(b) Let α and β be classifying maps for the linearly split extensions

$$0 \to \mathcal{C} \to \mathcal{E}_1 \to \mathcal{E}_2 \ldots \to \mathcal{E}_n \to \mathcal{B} \to 0$$

and

$$0 \to \mathcal{B} \to \mathcal{E}'_1 \to \mathcal{E}'_2 \ldots \to \mathcal{E}'_m \to \mathcal{A} \to 0$$

respectively. Then the classifying map for the Yoneda product of the two extensions is $\alpha \circ J^n(\beta) : J^{n+m}\mathcal{A} \to \mathcal{C}$.

Consider now the set $H_k = \langle J^k\mathcal{A}, \mathcal{K}\hat{\otimes}\mathcal{B}\rangle$, where $H_0 = \langle \mathcal{A}, \mathcal{K}\hat{\otimes}\mathcal{B}\rangle$. Each H_k is an abelian semigroup with 0-element for the K-theory addition defined in 4.2. Morally, the elements of H_k are classifying maps for linearly split k-step extensions. In applications all elements arise that way.

To define a map $S : H_k \to H_{k+2}$, we use the classifying map ε for the two-step extension which is obtained by composing the Toeplitz with the suspension extension, see B.4 and B.1.

Definition 4.5. *For each m-algebra $\mathcal{A}$, we define the periodicity map $\varepsilon_\mathcal{A} : J^2\mathcal{A} \to \mathcal{K}\hat{\otimes}\mathcal{A}$ as the classifying map for the standard two-step Bott extension*

$$0 \to \mathcal{K}\hat{\otimes}\mathcal{A} \to \mathcal{T}_0\hat{\otimes}\mathcal{A} \to \mathcal{A}[0,1) \to \mathcal{A} \to 0$$

defined as in Appendix B.4.

We can now define the Bott map S. For $\langle \alpha \rangle \in H_k$, $\alpha : J^k \mathcal{A} \longrightarrow \mathcal{K} \hat{\otimes} \mathcal{B}$ we set $S\langle \alpha \rangle = \langle (\mathrm{id}_\mathcal{K} \hat{\otimes} \alpha) \circ \varepsilon \rangle$. Here $\varepsilon : J^{k+2} \mathcal{A} \longrightarrow \mathcal{K} \otimes J^k \mathcal{A}$ is the ε-map for $J^k \mathcal{A}$.

If an element γ of H_k is given as a classifying map for an extension of length k, then $S\gamma$ is the classifying map for the Yoneda product of the given extension with the Bott extension.

Let $\varepsilon_- : J^{k+2} \mathcal{A} \longrightarrow \mathcal{K} \otimes J^k \mathcal{A}$ be the map which is obtained by replacing, in the definition of ε, the Toeplitz extension by the inverse Toeplitz extension. The sum $\varepsilon \oplus \varepsilon_-$ is then diffotopic to 0. Therefore $S\langle \alpha \rangle + S_-\langle \alpha \rangle = 0$, putting $S_-\langle \alpha \rangle = \langle (\mathrm{id}_\mathcal{K} \otimes \alpha) \circ \varepsilon_- \rangle$.

Definition 4.6. *Let $\mathcal{A}$ and $\mathcal{B}$ be m-algebras and $* = 0$ or 1. We define*

$$kk_*(\mathcal{A}, \mathcal{B}) = \varinjlim_k H_{2k+*} = \varinjlim_k \langle J^{2k+*} \mathcal{A}, \mathcal{K} \hat{\otimes} \mathcal{B} \rangle$$

The preceding discussion shows that $kk_*(\mathcal{A}, \mathcal{B})$ is not only an abelian semigroup, but even an abelian group (every element admits an inverse).

As usual for bivariant theories, the decisive element, which is also the most difficult to establish, is the composition product.

Theorem 4.7. *There is an associative product*

$$kk_i(\mathcal{A}, \mathcal{B}) \times kk_j(\mathcal{B}, \mathcal{C}) \longrightarrow kk_{i+j}(\mathcal{A}, \mathcal{C})$$

($i, j \in \mathbb{Z}/2$; $\mathcal{A}, \mathcal{B}$ and $\mathcal{C}$ m-algebras), which is additive in both variables.

Neglecting the tensor product by $\mathcal{K}$, the product of an element represented by $\varphi \in \langle J^k \mathcal{A}, \mathcal{K} \hat{\otimes} \mathcal{B} \rangle$ and an element represented by $\psi \in \langle J^l \mathcal{B}, \mathcal{K} \hat{\otimes} \mathcal{C} \rangle$ is defined as $\langle \psi \circ J^l(\varphi) \rangle$. For elements of kk which are given as classifying maps for higher length extensions, the product thus simply is the classifying map for the Yoneda product of the two extensions. The fact, that this is well defined, i.e. compatible with the periodicity map S, however demands a new idea, namely the "basic lemma" from [13].

Lemma 4.8. *Assume given a commutative diagram of the form*

$$\begin{array}{ccccccc}
& & 0 & & 0 & & 0 \\
& & \downarrow & & \downarrow & & \downarrow \\
0 \to & \mathcal{I} & \to & \mathcal{A}_{01} & \to & \mathcal{A}_{02} & \to 0 \\
& & \downarrow & & \downarrow & & \downarrow \\
0 \to & \mathcal{A}_{10} & \to & \mathcal{A}_{11} & \to & \mathcal{A}_{12} & \to 0 \\
& & \downarrow & & \downarrow & & \downarrow \\
0 \to & \mathcal{A}_{20} & \to & \mathcal{A}_{21} & \to & \mathcal{B} & \to 0 \\
& & \downarrow & & \downarrow & & \downarrow \\
& & 0 & & 0 & & 0
\end{array}$$

where all the rows and columns represent extensions of m-algebras with continuous linear splittings.

Let γ_+ and γ_- denote the classifying maps $J^2\mathcal{B} \to \mathcal{I}$ for the two extensions of length 2

$$0 \to \mathcal{I} \to \mathcal{A}_{10} \to \mathcal{A}_{21} \to \mathcal{B} \to 0$$

$$0 \to \mathcal{I} \to \mathcal{A}_{01} \to \mathcal{A}_{12} \to \mathcal{B} \to 0$$

associated with the two edges of the diagram. Then $\gamma_+ \oplus \gamma_-$ is diffotopic to 0.

This lemma implies that, for $\varphi : J^k\mathcal{A} \to \mathcal{B}$, the map $\varepsilon_\mathcal{B} \circ J^2(\varphi)$ is equivalent to $\varphi \circ J^k(\varepsilon_\mathcal{A})$ (again neglecting the tensor factor $\mathcal{K}$). For an extension

$$E: \quad 0 \to \mathcal{B} \to \mathcal{E}_1 \to \mathcal{E}_2 \ldots \to \mathcal{E}_k \to \mathcal{A} \to 0$$

this means that the classifying maps for the Yoneda products of E by the Bott extension for $\mathcal{B}$ on the left, and for the Yoneda product by the Bott extension for $\mathcal{A}$ on the right, are equivalent.

The following properties of kk can then be deduced in a rather standard fashion.

Theorem 4.9. *(a) There is a bilinear, graded commutative, exterior product*

$$kk_i(\mathcal{A}_1, \mathcal{A}_2) \times kk_j(\mathcal{B}_1, \mathcal{B}_2) \longrightarrow kk_{i+j}(\mathcal{A}_1\hat\otimes\mathcal{A}_2, \mathcal{B}_1\hat\otimes\mathcal{B}_2)$$

(b) Each homomorphism $\varphi : \mathcal{A} \to \mathcal{B}$ defines an element $kk(\varphi)$ in the group $kk_0(\mathcal{A}, \mathcal{B})$. If $\psi : \mathcal{B} \to \mathcal{C}$, is another homomorphism, then

$$kk(\psi \circ \varphi) = kk(\varphi) \cdot kk(\psi)$$

$kk_*(\mathcal{A}, \mathcal{B})$ is a contravariant functor in $\mathcal{A}$ and a covariant functor in $\mathcal{B}$. If $\alpha : \mathcal{A}' \to \mathcal{A}$ and $\beta : \mathcal{B} \to \mathcal{B}'$ are homomorphisms, then the induced maps, in the first and second variable of kk_*, are given by left multiplication by $kk(\alpha)$ and right multiplication by $kk(\beta)$.
(c) $kk_*(\mathcal{A}, \mathcal{A})$ is, for each m-algebra $\mathcal{A}$, a $\mathbb{Z}/2$-graded ring with unit element $kk(\mathrm{id}_\mathcal{A})$.
(d) The functor kk_* is invariant under diffotopies in both variables.
(e) The canonical inclusion $\iota : \mathcal{A} \to \mathcal{K}\hat\otimes\mathcal{A}$ defines an invertible element in $kk_0(\mathcal{A}, \mathcal{K}\hat\otimes\mathcal{A})$. In particular, $kk_*(\mathcal{A}, \mathcal{B}) \cong kk_*(\mathcal{K}\hat\otimes\mathcal{A}, \mathcal{B})$ and $kk_*(\mathcal{B}, \mathcal{A}) \cong kk_*(\mathcal{B}, \mathcal{K}\hat\otimes\mathcal{A})$ for each m-algebra $\mathcal{B}$.
(f) (Bott periodicity) The suspension extension gives rise to classifying maps $J\mathcal{A} \to \mathcal{A}(0,1)$ and $J^2\mathcal{A} \to \mathcal{A}(0,1)^2$ (where $\mathcal{A}(0,1)^2 = \mathcal{A}(0,1)(0,1)$). The elements in $kk_0(J\mathcal{A}, \mathcal{A}(0,1))$, in $kk_1(\mathcal{A}, \mathcal{A}(0,1))$ and in $kk_0(\mathcal{A}, \mathcal{A}(0,1)^2)$ defined by these homomorphisms are invertible.

Another essential property of the bivariant theory, which makes it computable, is the existence of long exact sequences associated with extensions of m-algebras.

Theorem 4.10. *Let $\mathcal{D}$ be any m-algebra. Every extension admitting a continuous linear section*

$$E: 0 \to \mathcal{I} \xrightarrow{i} \mathcal{A} \xrightarrow{q} \mathcal{B} \to 0$$

induces exact sequences in $kk(\mathcal{D}, \cdot)$ and $kk(\cdot, \mathcal{D})$ of the following form:

$$\begin{array}{ccccc}
kk_0(\mathcal{D}, \mathcal{I}) & \xrightarrow{\cdot kk(i)} & kk_0(\mathcal{D}, \mathcal{A}) & \xrightarrow{\cdot kk(q)} & kk_0(\mathcal{D}, \mathcal{B}) \\
\uparrow & & & & \downarrow \\
kk_1(\mathcal{D}, \mathcal{B}) & \xleftarrow{\cdot kk(q)} & kk_1(\mathcal{D}, \mathcal{A}) & \xleftarrow{\cdot kk(i)} & kk_1(\mathcal{D}, \mathcal{I})
\end{array} \quad (1)$$

and

$$\begin{array}{ccccc}
kk_0(\mathcal{I}, \mathcal{D}) & \xleftarrow{kk(i)\cdot} & kk_0(\mathcal{A}, \mathcal{D}) & \xleftarrow{kk(q)\cdot} & kk_0(\mathcal{B}, \mathcal{D}) \\
\downarrow & & & & \uparrow \\
kk_1(\mathcal{B}, \mathcal{D}) & \xrightarrow{kk(q)\cdot} & kk_1(\mathcal{A}, \mathcal{D}) & \xrightarrow{kk(i)\cdot} & kk_1(\mathcal{I}, \mathcal{D})
\end{array} \quad (2)$$

The given extension E defines a classifying map $J\mathcal{B} \to \mathcal{I}$ and thus an element of $kk_1(\mathcal{B}, \mathcal{I})$, which we denote by $kk(E)$. The vertical arrows in (1) and (2) are (up to a sign) given by right and left multiplication, respectively, by this class $kk(E)$.

The proof uses of course the composition product.

Finally, we have to compare kk with the ordinary topological K-theory. For a Fréchet m-algebra $\mathcal{A}$ one can define the following abelian groups:

$$K_0(\mathcal{A}) = \{ \langle e \rangle \,|\, e \text{ is an idempotent element in } M_2((\mathcal{K} \hat{\otimes} \mathcal{A})^\sim) \\ \text{such that } e - e_0 \text{ is in the ideal } M_2(\mathcal{K} \hat{\otimes} \mathcal{A})\} \quad (3)$$

$$K_1(\mathcal{A}) = K_0(\mathcal{A}(0,1))$$

Here, as usual $(\mathcal{K} \hat{\otimes} \mathcal{A})^\sim$ denotes the algebra obtained by adjoining a unit to $\mathcal{K} \hat{\otimes} \mathcal{A}$ and

$$e_0 = \begin{pmatrix} 1 & 0 \\ 0 & 0 \end{pmatrix}$$

Phillips shows in [41] that this defines a topological K-theory with the usual properties (homotopoy invariance, stability, Bott periodicity, six-term exact sequences) on the category of Fréchet m-algebras and in addition gives the usual well-known K-theory if $\mathcal{A}$ is a Banach algebra. We have

Theorem 4.11. *For every Fréchet m-algebra $\mathcal{A}$, the groups $kk_*(\mathbb{C}, \mathcal{A})$ and $K_*\mathcal{A}$ are naturally isomorphic.*

The proof of this theorem depends on the fact that topological K-theory has long exact sequences for extensions of Fréchet m-algebras and on the following alternative description of K-theory for such algebras (cf. also [11]).

Theorem 4.12. *For every Fréchet m-algebra $\mathcal{A}$, the group $K_0\mathcal{A}$ is isomorphic to $\langle q\mathbb{C}, \mathcal{K}\hat{\otimes}\mathcal{A}\rangle$, where $q\mathbb{C}$ is defined as in B.3.*

Remark 4.13. V.Lafforgue, [32] has introduced a bivariant K-theory KK^{Ban} which is defined on the category of Banach algebras. It does not have a product but it has very good properties for very general Morita equivalences of Banach algebras. Lafforgue uses his theory to obtain striking positive results on the Baum-Connes conjecture. For Banach algebras, it is possible to construct an analogue kk_*^{Ban} of kk_* on the category of Banach algebras using continuous homotopies rather than differentiable ones (this is straightforward) and to show that there is a natural transformation $KK_*^{\mathrm{Ban}} \to kk_*^{\mathrm{Ban}}$.

4.2 The bivariant Chern-Connes character

In this section we describe the construction of a bivariant multiplicative transformation from kk_* to the bivariant theory HP_* on the category of m-algebras. As a very special case it will furnish the correct frame for viewing the characters for idempotents, invertibles or Fredholm modules discussed above.

Consider a covariant functor E from the category of m-algebras to the category of abelian groups which satisfies the following conditions:

(E1) E is diffotopy invariant, i.e., the evaluation map ev_t in any point $t \in [0, 1]$ induces an isomorphism $E(ev_t) : E(\mathcal{A}[0,1]) \to E(\mathcal{A})$;
(E2) E is stable, i.e., the canonical inclusion $\iota : \mathcal{A} \to \mathcal{K}\hat{\otimes}\mathcal{A}$ induces an isomorphism $E(\iota) : E(\mathcal{A}) \to \mathcal{E}(\mathcal{K}\hat{\otimes}\mathcal{A})$;
(E3) E is half-exact, i.e., each extension $0 \to \mathcal{I} \to \mathcal{A} \to \mathcal{B} \to 0$ admitting a continuous linear splitting induces a short exact sequence $E(\mathcal{I}) \to E(\mathcal{A}) \to E(\mathcal{B})$.

(The same conditions can of course be formulated analogously for a contravariant functor E.) We note that a standard construction from algebraic topology, using property (E1) and mapping cones, permits to extend the short exact sequence in (E3) to an infinite long exact sequence of the form

$$\cdots \to E(\mathcal{B}(0,1)^2) \to E(\mathcal{I}(0,1)) \to E(\mathcal{A}(0,1))$$
$$\to E(\mathcal{B}(0,1)) \to E(\mathcal{I}) \to E(\mathcal{A}) \to E(\mathcal{B})$$

see, e.g., [28] or [1]. Most importantly for us here, however, it follows from (E1), (E2), (E3) that the periodicity map $\varepsilon : J^2\mathcal{A} \to \mathcal{K}\hat{\otimes}\mathcal{A}$ induces an isomrphism $E(\varepsilon) : E(J^2\mathcal{A}) \to E(\mathcal{K}\hat{\otimes}\mathcal{A})$.

Theorem 4.14. *Let E be a covariant functor with the properties (E1), (E2), (E3). Then we can associate in a unique way with each $h \in kk_0(\mathcal{A},\mathcal{B})$ a morphism of abelian groups $E(h) : E(\mathcal{A}) \to E(\mathcal{B})$, such that $E(h_1 \cdot h_2) = E(h_2) \circ E(h_1)$ for the product $h_1 \cdot h_2$ of $h_1 \in kk_0(\mathcal{A},\mathcal{B})$ and $h_2 \in kk_0(\mathcal{B},\mathcal{C})$, and such that $E(kk(\alpha)) = E(\alpha)$ for each morphism $\alpha : \mathcal{A} \to \mathcal{B}$ of m-algebras. An analogous statement holds for contravariant functors.*

Proof. Let h be represented by $\eta : J^{2n}\mathcal{A} \to \mathcal{K}\hat{\otimes}\mathcal{B}$. We set

$$E(h) = E(\iota)^{-1}E(\eta)E(\varepsilon^n)^{-1}E(\iota)$$

It is clear that $E(h)$ is well-defined and that $E(kk(\alpha)) = E(\alpha)$.

The preceding result can be interpreted differently, see also [1, 23]. For this, note first that kk_0 can be regarded as a category, whose objects are the m-algebras, and where the morphism set between $\mathcal{A}$ and $\mathcal{B}$ is given by $kk_0(\mathcal{A}, \mathcal{B})$. This category is additive in the sense that the morphism set between two objects forms an abelian group and that the product of morphisms is bilinear. We denote the natural functor from the category of m-algebras to the category kk_0, which is the identity on objects, also by kk_0.

Corollary 4.15. *Let F be a functor from the category of m-algebras to an additive category C, such that $F(\beta \circ \alpha) = F(\alpha) \cdot F(\beta)$, for any two homomorphisms $\alpha : A_1 \to A_2$ and $\beta : A_2 \to A_3$ between m-algebras.*

We assume that for each $\mathcal{B}$, the contravariant functor $C(F(\cdot), F(\mathcal{B}))$ and the covariant functor $C(F(\mathcal{B}), F(\cdot))$ on the category of m-algebras satisfy the properties (E1), (E2), (E3). Then there is a unique covariant functor F' from the category kk_0 to C, such that $F = F' \circ kk_0$.

Remark 4.16. Property (E3) implies that any such functor F' is automatically additive:

$$F'(h+g) = F'(h) + F'(g)$$

As a consequence of the preceding corollary we get a bilinear multiplicative transformation from kk_0 to HP_0 - the bivariant Chern-Connes character.

Corollary 4.17. *There is a unique (covariant) functor $ch : kk_0 \to HP_0$, such that $ch(kk_0(\alpha)) = HP_0(\alpha) \in HP_0(\mathcal{A}, \mathcal{B})$ for every morphism $\alpha : \mathcal{A} \to \mathcal{B}$ of m-algebras.*

Proof. This follows from 4.15, since HP_0 satisfies conditions (E1), (E2) and (E3) in both variables.

The Chern-Connes-character ch is by construction compatible with the composition product on kk_0. It also is compatible with the exterior product on kk_0, see 4.9(b), and the corresponding product on HP_0, see [17], p.86.

It remains to extend ch to a multiplicative transformation from the $\mathbb{Z}/2$-graded theory kk_* to HP_* and to study the compatibility with the boundary maps in the long exact sequences associated to an extension for kk and HP.

The natural route to the definition of ch in the odd case is the use of the identity $kk_1(\mathcal{A}, \mathcal{B}) = kk_0(J\mathcal{A}, \mathcal{B})$.

Since HP satisfies excision in the first variable and since $HP_*(T\mathcal{A}, \mathcal{B}) = 0$ for all $\mathcal{B}$ ($T\mathcal{A}$ is contractible), we find that

$$HP_0(J\mathcal{A}, \mathcal{B}) \cong HP_1(\mathcal{A}, \mathcal{B}) \tag{4}$$

However, for the product $kk_1 \times kk_1 \to kk_0$ we have to use the identification
$$kk_0(J^2\mathcal{A},\mathcal{B}) \cong kk_0(\mathcal{A},\mathcal{B})$$
which is induced by the ε-map $J^2\mathcal{A} \to \mathcal{K}\hat{\otimes}\mathcal{A}$. This identification is different from the identification $HP_0(J^2\mathcal{A},\mathcal{B}) \cong HP_0(\mathcal{A},\mathcal{B})$ which we obtain by applying (4) twice. In fact, we have

Proposition 4.18. *Under the natural identification $HP_0(J^2\mathcal{A},\mathcal{K}\hat{\otimes}\mathcal{A}) \cong HP_0(\mathcal{A},\mathcal{A})$ from 4, the element $ch(\varepsilon)$ corresponds to $(2\pi i)^{-1}1_\mathcal{A}$.*

We are therefore lead to the following definition.

Definition 4.19. *Let u be an element in $kk_1(\mathcal{A},\mathcal{B})$ and let u_0 be the corresponding element in $kk_0(J\mathcal{A},\mathcal{B})$. We set*
$$ch(u) = \sqrt{2\pi i}\; ch(u_0) \quad \in HP_1(\mathcal{A},\mathcal{B}) \cong HP_0(J\mathcal{A},\mathcal{B})$$

Theorem 4.20. *The thus defined Chern-Connes character $ch : kk_* \to HP_*$ is multiplicative, i.e., for $u \in kk_i(\mathcal{A},\mathcal{B})$ and $v \in kk_j(\mathcal{B},\mathcal{C})$ we have*
$$ch(u \cdot v) = ch(u) \cdot ch(v)$$

It follows that the character is also compatible with the boundary maps in the six-term exact sequences induced by an extension in both variables of kk_* and HP_*.

For m-algebras, the bivariant character ch constructed here is a far reaching generalization of the Chern character from K-Theory and K-homology in Sects. 2.10 and 3.4.

If e is an idempotent in $M_n\mathcal{A}$, then e can be viewed as a homomorphism from $\mathbb{C}$ to $\mathcal{K}\otimes\mathcal{A}$, thus as an element h_e of $kk_0(\mathbb{C},\mathcal{A})$. It is not difficult to see that for the class $ch([e])$ constructed in section 2.10 one has
$$ch([e]) = ch(h_e)$$
Similarly any invertible element u in $M_n\mathcal{A}^\sim$ defines an element h_u in $kk_1(\mathbb{C},\mathcal{A})$ and we have $ch([u]) = ch(h_u)$

Finally, any p-summable Fredholm module over $\mathcal{A}$ gives an element k of $kk_*(\mathcal{A},\ell^p(H))$.

Corollary 3.4 shows that $\ell^p\hat{\otimes}\mathcal{B}$ and $\ell^1\hat{\otimes}\mathcal{B}$ are equivalent in HP_0 for each $p \geq 1$. On the other hand, it is easy to see that $\ell^1\hat{\otimes}\mathcal{B}$ is in HP_0 equivalent to $\mathcal{B}$, see for instance [19]. We thus obtain

Corollary 4.21. *The m-algebra $\ell^p\hat{\otimes}\mathcal{B}$ is HP-equivalent to $\mathcal{B}$ for each $p \geq 1$. The Chern-Connes-character gives a transformation $ch^{(p)}: kk_*(\mathcal{A},\ell^p\hat{\otimes}\mathcal{B}) \to HP_*(\mathcal{A},\mathcal{B})$ such that*
$$ch^{(p)}(x \cdot kk(\iota^{(p)})) = ch(x) \qquad \text{for } x \in kk_*(\mathcal{A},\mathcal{B})$$
where $\iota^{(p)}$ denotes the canonical inclusion $\mathcal{B} \to \ell^p\hat{\otimes}\mathcal{B}$.

Comparing the functorial properties, one sees that this transformation generalizes the construction of cyclic cohomology classes associated to p-summable Fredholm modules in Sect. 3.4. It also generalizes previous constructions of a bivariant Chern-Connes character in special cases, [30, 38, 39, 52].

5 Infinite-dimensional Cyclic Theories

5.1 Entire cyclic cohomology

In the study of discrete subgroups of Lie groups and of quantum field theory, one is naturally lead to consider Fredholm modules that are not finitely summable but only θ-summable:

Definition 5.1. *A Fredholm module (F, π, H) is called θ-summable iff the (graded) commutators $[F, \pi(a)]$ are contained in the operator ideal*

$$\mathrm{Li}^{1/2}(H) := \{T \in \mathcal{L}(H) \mid \mu_n(T) = O\big((\ln n)^{-1/2}\big) \text{ for } n \to \infty\}.$$

Here $\mu_n(T)$ denotes the nth singular value of T.

In periodic cyclic cohomology, one can only express the Chern character of a finitely summable Fredholm module. Entire cyclic cohomology was introduced to handle θ-summable Fredholm modules as well [5].

Definition 5.2. *Let A be a Banach algebra. Let K be the $\mathbb{Z}/2$-graded vector space of all linear maps $\phi\colon \Omega A \to \mathbb{C}$ that satisfy the following growth condition: For all bounded subsets $S \subseteq A$, there is a constant C_S such that*

$$|\phi(a_0 da_1 \ldots da_n)| \leq C_S, \qquad |\phi(da_1 \ldots da_n)| \leq C_S \qquad (1)$$

for all $a_0, \ldots, a_n \in S$. We equip K with the boundary $\partial^*\colon K \to K$ induced from the boundary maps of $X(TA)$. That is, $\partial^*(\phi) := \phi \circ \beta$ for even ϕ and $\partial^*(\phi) := -\phi \circ \delta$ for odd ϕ, see 2.29. The entire cyclic cohomology $HE^*(A)$ is defined to be the homology of the complex K.

If we use $B \pm b$ instead of the X-complex boundary, we have to modify the growth condition above, replacing (1) by

$$[n/2]! \cdot |\phi(a_0 da_1 \ldots da_n)| \leq C_S, \qquad [n/2]! \cdot |\phi(da_1 \ldots da_n)| \leq C_S, \qquad (2)$$

where $[n/2] = k$ if $n = 2k$ or $n = 2k+1$. This yields a chain homotopic complex [35].

Definition 5.2 can be carried over literally to locally convex algebras with separately continuous multiplication. However, since we only use the bounded subsets of A, it is more appropriate to define entire cyclic cohomology for *complete (convex) bornological algebras*. These are algebras equipped with a collection $\mathcal{S}$ of subsets satisfying some reasonable axioms like

$$S, T \in \mathcal{S}, \ \lambda \in \mathbb{C} \Longrightarrow S \cdot T, \ S + T, \ \lambda \cdot S \in \mathcal{S}$$

Convexity means that if $S \in \mathcal{S}$, so is the absolutely convex hull of S. If $S \subseteq A$ is absolutely convex, then the linear span A_S of S carries a unique semi-norm with unit ball S. We call S *completant* iff A_S is a Banach space. We call A *separated* iff A_S is a normed space for all S and we call A *complete* iff each $S \in \mathcal{S}$ is contained in a completant absolutely convex set $T \in \mathcal{S}$. A complete bornological vector space is automatically separated. In the following, we will drop the words convex and separated from our notation and assume all bornological vector spaces to be convex and separated. The collection $\mathcal{S}$ is called the *bornology* of A. We call $S \subseteq A$ *small* iff $S \in \mathcal{S}$. See [24] for an introduction to bornological vector spaces.

Here are some important examples of bornologies. If V is just a vector space, we can equip it with the finest possible bornology: We let $S \in \mathcal{S}$ iff S is a bounded subset of a finite dimensional vector subspace of V. If V is a (quasi)complete locally convex vector space, then the collection of bounded subsets of V is a complete bornology. Another reasonable complete bornology is the collection of all precompact or relatively compact subsets. (If V is (quasi)complete, then a subset is precompact iff it is relatively compact iff it is contained in a compact subset.) Actually, the precompact bornology tends to be better behaved than the bounded one. For local cyclic cohomology, it is essential to use the precompact bornology.

Let A be a bornological algebra. Then there is an obvious bornology on ΩA such that the space K of Definition 5.2 is equal to the space of bounded linear functionals on ΩA. Namely, the bornology generated by the sets $\bigcup_{n \geq 0} S(dS)^n \cup (dS)^{n+1}$. That is, a subset of ΩA is small if and only if it is contained in the absolutely convex hull of a set of the form $\bigcup_{n \geq 0} S(dS)^n \cup (dS)^{n+1}$. The operators b, d, B, κ and the multiplication of differential forms are bounded with respect to this bornology. Identifying TA and $X(TA)$ with $\Omega^{\text{ev}} A$ and ΩA as in Proposition 2.27, we obtain canonical bornologies on TA and $X(TA)$. We always equip $\Omega A \cong X(TA)$ and TA with these bornologies.

If V is a bornological vector space, we write V^c for its *completion*. Moreover, we write $\mathcal{L}(X, Y)$ for the space of bounded linear maps between two bornological vector spaces X and Y.

Definition 5.3. *Let A and B be bornological algebras. We define* entire cyclic homology *and* bivariant entire cyclic homology *by*

$$HE_*(B) := H_*\big(X(TB)^c\big),$$
$$HE_*(A, B) := H_*\Big(\mathcal{L}\big(X(TA)^c, X(TB)^c\big)\Big).$$

It is necessary to complete $X(TB)$ because the complex $X(TB)$ itself is always acyclic. We have $HE_*(A, \mathbb{C}) \cong HE^*(A)$ and $HE_*(\mathbb{C}, A) \cong HE_*(A)$ because the complex $X(T\mathbb{C})^c$ is chain homotopic to the complex $\mathbb{C}$ with $\mathbb{C}$ in degree 0 and 0 in degree 1.

The composition of chain maps gives rise to an associative *composition product* in bivariant entire cyclic homology. A bounded homomorphism $f\colon A \to B$ induces a bounded chain map $f_*\colon X(TA)^c \to X(TB)^c$. Thus we obtain an element $[f_*] \in HE_0(A,B)$. Composition with $[f_*]$ yields the functoriality of the entire cyclic theories.

Since $X(TA)^c$ is contained in the complex $\hat{X}(TA)$ defining the periodic cyclic theories, there are natural transformations $HP^*(A) \to HE^*(A)$ and $HE_*(A) \to HP_*(A)$.

Entire cyclic (co)homology has the same stability, homotopy invariance, and exactness properties as periodic cyclic (co)homology [35].

To formulate homotopy invariance and stability, we need the completed bornological tensor product $\hat{\otimes}$. It is defined by a universal property analogous to the universal property of the projective tensor product: bounded linear maps $V \hat{\otimes} W \to X$ correspond bijectively to bounded bilinear maps $V \times W \to X$. For Banach spaces equipped with the bounded bornology, $V \hat{\otimes} W$ is nothing but the projective tensor product, again equipped with the bounded bornology. If V and W are Fréchet spaces equipped with the precompact bornology, then $V \hat{\otimes} W$ will be isomorphic to the projective tensor product $V \hat{\otimes}_\pi W$ equipped with the precompact bornology [35]. A slightly weaker statement holds if V and W carry the bounded bornology.

For a complete bornological algebra, we define $B[0,1] := \mathbb{C}[0,1] \hat{\otimes} B$, where $\mathbb{C}[0,1]$ carries the bounded bornology (which is equal to the precompact bornology). This gives rise to a notion of smooth homotopy. Entire cyclic (co)homology is smoothly homotopy invariant, that is, Theorem 2.45.(a) holds with HE instead of HP. Similarly, entire cyclic homology is stable with respect to the trace class operators $\ell^1(H)$. That is, the canonical embedding $A \to A \hat{\otimes} \ell^1(H)$ is invertible in HE_0. These two assertions can be proved as for the periodic theory. The excision theorem also holds [35, 44]:

Theorem 5.4. *Bivariant entire cyclic homology satisfies excision in both variables for extensions with a bounded linear section.*

That is, if $0 \to S \to P \to Q \to 0$ is an extension of complete bornological algebras with a bounded linear section and if A is another complete bornological algebra, then there are two natural six-term exact sequences as in Theorem 2.50, only with HE instead of HP. However, the proof of Theorem 5.4 is more difficult than the proof of Theorem 2.50. We will explain the strategy of the proof below.

The formulas for the Chern character in K-theory in Sect. 2.10 actually yield elements of $X(TA)^c$. Hence these formulas define a Chern character

$$\mathrm{ch}\colon K_*(A) \to HE_*(A).$$

The Chern character $K_*(A) \to HP_*(A)$ is the composition of this character with the natural map $HE_*(A) \to HP_*(A)$.

There are two apparently different definitions for the Chern character of a θ-summable Fredholm module, due to Connes [5] and Jaffe-Lesniewski-Osterwalder [25], respectively. It is an important fact that both approaches yield homologous cocycles [6]. In [25], an entire cochain is written down explicitly and it is checked that it is a cocycle. However, it is not clear why this particular formula should be the right one. Connes's construction is more conceptual. Presumably, it can be explained using the excision theorem, as in Sect. 3.4. However, Connes's construction is technically quite involved, so that we cannot discuss it here.

Both constructions of the Chern character work with *unbounded* Fredholm modules in the following sense.

Definition 5.5. *Let A be a bornological algebra. Let H be a Hilbert space, let $\pi\colon A \to \mathcal{L}(H)$ be a bounded representation of A, and let D be an unbounded self-adjoint operator on H. We call the triple (D, π, H) an* unbounded odd Fredholm *module over A iff $(1 + D^2)^{-1}$ is compact, $D\pi(a)$ is densely defined for all $a \in A$, $[D, \pi(a)]$ extends to a bounded operator on H for all $a \in A$, and the map $A \ni a \mapsto [D, \pi(a)] \in \mathcal{L}(H)$ is bounded.*

If H is a graded Hilbert space, $\pi(a)$ has degree 0 for all $a \in A$, and D has degree 1, we call the triple (D, π, H) an unbounded even Fredholm module.

We call (D, π, H) θ-summable iff $\exp(-D^2) \in \ell^1(H)$.

In applications, it frequently happens that $\exp(-\beta D^2) \in \ell^1(H)$ for all $\beta > 0$. This stronger assumption is made in [5] and [25], although it is not necessary to obtain the Chern character.

Given an unbounded Fredholm module (D, π, H), we get an associated bounded Fredholm module as follows. If D is injective, we may put $F := \operatorname{sign} D$. Otherwise, we first modify (D, π, H) so that D becomes injective and then put $F := \operatorname{sign} D$. In the odd case, we may simply add to D the projection onto the kernel of D. In the even case, this does not work because the resulting operator will no longer have degree 1. In fact, there may be no injective self-adjoint degree 1 operator on $\ker D$. Therefore, we replace H by the direct sum $H \oplus \ker D$ and extend $\pi(a)$ and D to be 0 on the summand $\ker D$. We equip the summand $\ker D$ with the opposite grading. This insures that the operator

$$\begin{pmatrix} 0 & 1 \\ 1 & 0 \end{pmatrix}$$

on $\ker D \oplus \ker D$ has degree 1. Adding this operator to $D \oplus 0$, we get an unbounded Fredholm module $(D', \pi \oplus 0, H \oplus \ker D)$ with $\ker D' = \{0\}$.

If we start with a θ-summable unbounded Fredholm module in the sense of Definition 5.5, the resulting bounded Fredholm module will be θ-summable as well [47]. Conversely, if A is countably generated, then any bounded θ-summable Fredholm module arises in this way from a suitable unbounded θ-summable Fredholm module [7]. This construction only yields $\exp(-D^2) \in \ell^1(H)$, not $\exp(-\beta D^2) \in \ell^1(H)$ for all $\beta > 0$.

Let (D, π, H) be an unbounded, even, θ-summable Fredholm module with grading operator γ. The associated *JLO-cocycle* is defined as follows. Let

$$\Delta_k := \{(s_0, \ldots, s_k) \in [0,1]^{k+1} \mid s_0 + \cdots + s_k = 1\}$$

be the standard k-simplex and let $\int_{\Delta_k}$ denote integration over Δ_k with respect to the usual Lebesgue measure $ds_0 \ldots ds_{k-1}$. The JLO-cocycle is defined by the formula

$$\tau(a_0 da_1 \ldots da_{2n}) := \int_{\Delta_{2n}} \text{Tr}\left(\gamma a_0 e^{-s_0 D^2}[D, a_1]e^{-s_1 D^2} \cdots [D, a_{2n}]e^{-s_{2n} D^2}\right) \tag{3}$$

for $n \geq 0$. An involved computation shows that $\tau \circ (B+b) = 0$ and that τ satisfies the growth condition (2). Hence τ defines an element of $HE^0(A)$. For odd Fredholm modules, almost the same formula works, see [7]. Recently, D. Perrot, [40], has constructed a bivariant generalization of the JLO-cocycle using Quillen's cochain formalism.

As expected, the Chern character for Fredhom modules is compatible with the Chern character in K-theory ch: $K_*(A) \to HE_*(A)$. That is, the analogues of Theorem 3.9 and 3.10 hold:

Theorem 5.6. *Let $x \in K_*(A)$ and let y be a θ-summable Fredholm module with the same parity as x. Then the pairing between the Chern characters $\langle \text{ch}\, x, \text{ch}\, y \rangle$ equals the index pairing between K-theory and K-homology.*

Viewing entire cyclic cohomology as a version of periodic cyclic cohomology that contains, in addition, certain infinite dimensional cycles, we expect that $HE^*(A) = HP^*(A)$ for "finite dimensional" algebras. A positive result in this direction is the following [31]:

Theorem 5.7. *Let A be a Banach algebra of finite Hochschild cohomological dimension. Then $HE^*(A) \cong HP^*(A)$.*

However, there are quasi-free Fréchet algebras for which entire and periodic cyclic cohomology differ.

Entire and periodic cyclic cohomology have many properties in common. An explanation for this is that we can describe both $HP^0(A)$ and $HE^0(A)$ using traces on "nilpotent" extensions of A. For the purposes of periodic cyclic cohomology, a nilpotent extension of A is an extension of algebras $N \to E \to A$ in which the kernel N is nilpotent in the sense that $N^n = 0$ for some $n \in \mathbb{N}$. A trace on E gives rise to an element of $HP^0(E)$ and hence of $HP^0(A)$ because Goodwillie's theorem asserts that $HP^0(A) \cong HP^0(E)$, see [20]. Moreover, all elements of $HP^0(A)$ arise in this way. It is also possible to describe when two traces give rise to the same element in $HP^0(A)$, but we will not explain this here.

We get entire cyclic cohomology by weakening the notion of nilpotence.

Definition 5.8. *A bornological algebra A is called* analytically nilpotent *iff $S^\infty := \bigcup_{n \geq 1} S^n$ is small for all $S \in \mathcal{S}(A)$. An* analytically nilpotent extension *of A is an extension $N \to E \to A$ of complete bornological algebras with a bounded linear section and analytically nilpotent kernel N.*

Let A be a complete bornological algebra that is analytically nilpotent and let $a \in A$. If the power series $\sum c_j z^j$ has non-zero radius of convergence, then the series $\sum c_j a^j$ converges in A. This means that all elements of A have "spectral radius zero". A typical example of an analytically nilpotent algebra is JA or its completion $(JA)^c$. Indeed, the bornology on TA is the coarsest one for which JA is analytically nilpotent and the canonical linear map $A \to TA$ is bounded.

Consequently, the extension $(JA)^c \to (TA)^c \to A$ is universal among the analytically nilpotent extensions in the following sense: If $N \to E \to A$ is an analytically nilpotent extension of A, then there is a morphism of extensions from $(TA)^c \to A$ to $E \to A$, that is, a commutative diagram

$$\begin{array}{ccccc} (JA)^c & \longrightarrow & (TA)^c & \longrightarrow & A \\ \downarrow & & \downarrow & & \| \\ N & \longrightarrow & E & \longrightarrow & A \end{array}$$

In addition, this morphism is unique up to smooth homotopy.

Definition 5.9. *Let $l : A \to B$ be a linear map between bornological algebras. For $S \in \mathcal{S}(A)$, let*

$$\omega_l(S, S) := \{l(xy) - l(x)l(y) \mid x, y \in S\} \subseteq B.$$

We say that l is almost multiplicative *or that l has* analytically nilpotent curvature *iff $\bigcup_{n \geq 1} \omega_l(S, S)^n$ is small for all $S \in \mathcal{S}(A)$.*

It is an important fact that almost multiplicative linear maps form a category, that is, compositions of almost multiplicative linear maps are again almost multiplicative. This is closely related to the observation that the class of analytically nilpotent algebras is closed under extensions. That is, if $A \to B \to C$ is an extension with analytically nilpotent A and C, then B is analytically nilpotent as well. Working with almost multiplicative linear maps, it is easy to obtain the following analogue of Goodwillie's theorem:

Theorem 5.10. *Let $N \to E \xrightarrow{p} A$ be an extension of complete bornological algebras with a bounded linear section. Suppose that N is analytically nilpotent. Then $[p_*] \in HE_0(E, A)$ is invertible. Therefore, $p^* : HE^0(A) \to HE^0(E)$ is an isomorphism.*

Another consequence of the composability of almost multiplicative maps is that any extension $N \to E \to (TA)^c$ with analytically nilpotent kernel N splits by a bounded homomorphism. Thus $(TA)^c$ is analytically quasi-free:

Definition 5.11. *A complete bornological algebra R is called* analytically quasi-free *iff any analytically nilpotent extension $N \to E \to R$ splits by a bounded homomorphism.*

Analytical quasi-freeness is a much stronger condition than quasi-freeness and is not related to homological algebra. Examples of analytically quasi-free algebras besides $(TA)^c$ are $\mathbb{C}$ and the algebra of Laurent series $\mathbb{C}[u, u^{-1}]$, equipped with the finest possible bornology.

Theorem 5.12. *Let $N \to R \to A$ be an extension of complete bornological algebras with a bounded linear section. If N is analytically nilpotent and R is analytically quasi-free, then the complexes $X(R)^c$ and $X(TA)^c$ are homotopy equivalent via bounded chain maps and homotopies.*

We now sketch the proof of the Excision Theorem 5.4 in [36]. Let $S \to P \to Q$ be an extension with bounded linear section. The basic idea of the proof is to play with the *left* ideal $L \subseteq TP$ that is generated by $S \subseteq TP$. It turns out that while the two-sided ideal generated by S does not have good properties, L behaves very nicely. First, there is an explicit free L-bimodule resolution of $\tilde{L}$ of length 1 related to the embedding $L \subseteq TP$. It follows that L is quasi-free. An analysis of this resolution shows that the complex $X(L)$ is homotopy equivalent to the kernel of the canonical map $X(TP) \to X(TQ)$. Secondly, L is not just quasi-free but analytically quasi-free. This is more difficult to show. One can construct a bounded homomorphism $L \to TL$ using the universal property of TP. The kernel of the canonical projection $L \to S$ is contained in JP and therefore analytically nilpotent. Hence Theorem 5.12 shows that $X(TS)$ and $X(L)$ are homotopy equivalent. Since $X(L)$ is homotopy equivalent to the kernel of the canonical map $X(TP) \to X(TQ)$, the Excision Theorem follows.

5.2 Local cyclic cohomology

Neither periodic nor entire cyclic cohomology yield good results for big algebras like C^*-algebras. If A is a nuclear C^*-algebra or, more generally, an amenable Banach algebra, then $HE^1(A) = HP^1(A) = 0$ and $HE^0(A) \cong HP^0(A) \cong HH^0(A)$ is the space of traces on A [31]. To obtain reasonable results for C^*-algebras as well, one has to use the local cyclic cohomology defined in [43].

The definition of local cyclic cohomology is based on entire cyclic cohomology. The first modification with respect to the entire theory is that we equip a Banach (or Fréchet) algebra A with the bornology of all *precompact* subsets. This change of bornology can still be handled within the framework of section 5.1. The second modification is that we view a bornological vector space as an inductive system and replace the ordinary cohomology by a "local" cohomology that is defined in the setting of inductive systems.

Let X be a bornological vector space. Then the absolutely convex small subsets of X form a directed set with respect to inclusion. Recall that if $S \subseteq X$

is an absolutely convex small subset, then its linear span X_S carries a canonical norm. The map $S \mapsto X_S$ defines an inductive system of normed spaces with the additional property that all the structure maps are injective. Its limit (in the category of bornological vector spaces) is isomorphic to X. Thus any bornological vector space can be written in a canonical way as an inductive limit of normed spaces. This construction gives rise to an equivalence between the category of bornological vector spaces and bounded linear maps and the category of inductive systems of normed spaces with *injective* structure maps and morphisms of inductive systems [35].

If we apply this construction to a $\mathbb{Z}/2$-graded complex of bornological vector spaces, the resulting inductive system is (isomorphic to) an inductive system of $\mathbb{Z}/2$-graded complexes of normed spaces. In the following, we view $X(TA)$ as an inductive system of $\mathbb{Z}/2$-graded complexes of normed spaces in this way. *We still write $X(TA)$ for this inductive system to simplify notation.*

The space of bounded linear functionals on an inductive system $(C_i)_{i \in I}$ is equal to the projective limit of the dual spaces C'_i. The projective limit functor has the following undesirable property: even if all the complexes C'_i are contractible, it may happen that the cohomology of the inductive limit is non-zero. We would like to have another cohomology functor H^*_{loc} that is *local* in the following sense: if $H^*(C_i) = 0$ for all $i \in I$, then $H^*_{\mathrm{loc}}((C_i)_{i \in I}) = 0$ as well.

To define H^*_{loc}, we observe the following. The category $\mathrm{Ho}(\mathrm{Ind})$ of inductive systems of $\mathbb{Z}/2$-graded complexes of normed spaces with homotopy classes of chain maps as morphisms is a triangulated category. The subclass of inductive systems of *contractible* complexes generates a "null system" $\mathcal{N}$ in $\mathrm{Ho}(\mathrm{Ind})$. Thus the corresponding quotient category $\mathrm{Ho}(\mathrm{Ind})/\mathcal{N}$ is again a triangulated category. The decisive property of this quotient category is that in it all elements of $\mathcal{N}$ become isomorphic to the zero object.

In the following definition, we view $\mathbb{C}$ as a constant inductive system of complexes with zero boundary as usual.

Definition 5.13. *Let $K = (K_i)_{i \in I}$ and $L = (L_j)_{j \in J}$ be two inductive systems of complexes. Define $H^{\mathrm{loc}}_*(K, L)$ to be the space of morphisms $K \to L$ in the category $\mathrm{Ho}(\mathrm{Ind})/\mathcal{N}$. Let*

$$H^*_{\mathrm{loc}}(K) := H^{\mathrm{loc}}_*(K, \mathbb{C}), \qquad H^{\mathrm{loc}}_*(L) := H^{\mathrm{loc}}_*(\mathbb{C}, L).$$

One can compute $H^{\mathrm{loc}}_*(K, L)$ via an appropriate projective resolution of K (see [43]). An analysis of this resolution yields that $H^{\mathrm{loc}}_*(K, L)$ is related to a spectral sequence whose E^2-term involves the homologies $H_*(\mathcal{L}(K_i, L_j))$ and the derived functors of the projective limit functor $\varprojlim$. For countable inductive systems, the derived functors $R^p \varprojlim$ with $p \geq 2$ vanish. Hence the spectral sequence degenerates to a Milnor $\varprojlim^1$-exact sequence

$$0 \to \varprojlim{}^1_I \varinjlim_J H_{*-1}\bigl(\mathcal{L}(K_i, L_j)\bigr)$$
$$\to H^{\mathrm{loc}}_*(K, L) \to \varprojlim_I \varinjlim_J H_*\bigl(\mathcal{L}(K_i, L_j)\bigr) \to 0$$

if I is countable. In the local theory we have this exact sequence for arbitrary countable inductive systems (K_i). In particular, for $K = \mathbb{C}$ we obtain

$$H_*^{\mathrm{loc}}(L) = \varinjlim H_*(L_j).$$

Since the inductive limit functor is exact, $H_*^{\mathrm{loc}}(L)$ is equal to the homology of the inductive limit of L.

The completion of an inductive system of normed spaces $(X_i)_{i \in I}$ is defined entry-wise: $(X_i)^{\mathrm{c}} := (X_i^{\mathrm{c}})_{i \in I}$.

Definition 5.14. *Let A and B be complete bornological algebras. We define* local cyclic cohomology, local cyclic homology, *and* bivariant local cyclic homology *by*

$$HE^*_{\mathrm{loc}}(A) := H^*_{\mathrm{loc}}\big(X(TA)\big), \qquad HE^{\mathrm{loc}}_*(B) := H^{\mathrm{loc}}_*\big(X(TB)^{\mathrm{c}}\big),$$
$$HE^{\mathrm{loc}}_*(A,B) := H^{\mathrm{loc}}_*\big(X(TA)^{\mathrm{c}}, X(TB)^{\mathrm{c}}\big) \cong H^{\mathrm{loc}}_*\big(X(TA), X(TB)^{\mathrm{c}}\big).$$

As usual, we have $HE^{\mathrm{loc}}_*(A, \mathbb{C}) \cong HE^*_{\mathrm{loc}}(A)$ and $HE^{\mathrm{loc}}_*(\mathbb{C}, A) \cong HE^{\mathrm{loc}}_*(A)$. The composition of morphisms gives rise to a product on HE^{loc}_*.

If we view a bornological vector space X as an inductive system (X_i) with injective structure maps, then the structure maps of the completion $(X_i)^{\mathrm{c}}$ need not be injective any more, so that $(X_i)^{\mathrm{c}}$ may differ from the inductive system that corresponds to the bornological vector space X^{c}. Hence we have to be careful in what category we form the completion $X(TB)^{\mathrm{c}}$. For the local theory, it is crucial to take the completion in the category of inductive systems. A consequence is that, in general, $HE^{\mathrm{loc}}_*(B)$ may be different from $HE_*(B)$. However, in applications it often turns out that the two completions coincide. This holds, for instance, if B is a Fréchet algebra equipped with the precompact bornology or if B is nuclear in the sense of Grothendieck. In these two cases we have $HE_*(B) \cong HE^{\mathrm{loc}}_*(B)$ and a natural map $HE_*(A,B) \to HE^{\mathrm{loc}}_*(A,B)$.

The main advantage of local cyclic cohomology is that it behaves well when passing to "smooth" subalgebras.

Definition 5.15. *Let A be a complete bornological algebra with bornology $\mathcal{S}(A)$, let B be a Banach algebra with closed unit ball U, and let $j \colon A \to B$ be an injective bounded homomorphism with dense image. We call A a* smooth subalgebra *of B iff $S^{\infty} := \bigcup S^n \in \mathcal{S}(A)$ whenever $S \in \mathcal{S}(A)$ and $j(S) \subseteq rU$ for some $r < 1$.*

Let $A \subseteq B$ be a smooth subalgebra. Then an element $a \in A$ that is invertible in B is already invertible in A. Hence A is closed under holomorphic functional calculus. In the setting of topological algebras, smooth subalgebras are defined in [2] using norms with special properties. If $A \subseteq B$ is a smooth subalgebra in the sense of [2], then A equipped with the precompact bornology is a smooth subalgebra in the sense of Definition 5.15.

Whereas the inclusion of a smooth subalgebra $A \to B$ induces an isomorphism $K_*(A) \cong K_*(B)$ in K-theory, the periodic or entire cyclic theories of A and B may differ drastically. However, the local theory behaves like K-theory in this situation:

Theorem 5.16. *Let $A \subseteq B$ be a smooth subalgebra. Assume that A and B are both endowed with the bornology of precompact subsets. Let $j\colon A \to B$ be the inclusion and let $j_* \in HE_0^{\mathrm{loc}}(A,B)$ be the corresponding element in the bivariant local cyclic homology.*

If B has Grothendieck's approximation property, then j_* is invertible. More generally, j_* is invertible if the identity map on B can be approximated uniformly on small subsets of B by maps of the form $j \circ l$ with $l\colon B \to A$ bounded and linear.

Bivariant local cyclic homology is exact for extensions with a bounded linear section, homotopy invariant for smooth homotopies and stable with respect to tensor products with the trace class operators $\ell^1(H)$. The proofs are the same as for entire cyclic cohomology. Using Theorem 5.16, we can strengthen these properties considerably:

Theorem 5.17. *Let A be a C^*-algebra. The functors $B \mapsto HE_*^{\mathrm{loc}}(A,B)$ and $B \mapsto HE_*^{\mathrm{loc}}(B,A)$ are split exact, stable homotopy functors on the category of C^*-algebras.*

For separable C^*-algebras, there is a natural bivariant Chern character

$$\mathrm{ch}\colon KK_*(A,B) \to HE_*^{\mathrm{loc}}(A,B).$$

The Chern character is multiplicative with respect to the Kasparov product on the left and the composition product on the right hand side.

If both A and B satisfy the Universal Coefficient Theorem in Kasparov theory, then there is a natural isomorphism

$$HE_*^{\mathrm{loc}}(A,B) \cong \mathrm{Hom}\bigl(K_*(A) \otimes_{\mathbb{Z}} \mathbb{C}, K_*(B) \otimes_{\mathbb{Z}} \mathbb{C}\bigr).$$

Proof. Since $C^\infty([0,1],A) \subseteq C([0,1],A)$ is a smooth subalgebra, Theorem 5.16 and smooth homotopy invariance imply continuous homotopy invariance. The subalgebra $M_\infty(A) = M_\infty \hat\otimes A$ defined as in (1) is a smooth subalgebra of the C^*-algebraic stabilization of A. Hence Theorem 5.16 and stability with respect to M_∞ imply C^*-algebraic stability.

The existence of the bivariant Chern character follows from these homological properties by the universal property of Kasparov's KK-theory.

The last assertion is trivial for $A = B = \mathbb{C}$. The class of C^*-algebras for which it holds is closed under KK-equivalence, inductive limits (with approximation property) and extensions with completely positive section. Hence it contains all nuclear C^*-algebras satisfying the Universal Coefficient Theorem (see [1]).

Examples of C^*-algebras satisfying the universal coefficient theorem are commutative C^*-algebras and the C^*-algebras of amenable groupoids [51].

Assume that Γ is the fundamental group of a compact Riemannian manifold M of non-positive sectional curvature or that Γ is a torsion-free word-hyperbolic group in the sense of Gromov. Then Γ has a finite model, so that $\mathbb{C}[\Gamma]$ has finite homological dimension. We have [45]

$$HE_*^{\mathrm{loc}}(\ell^1(\Gamma)) \cong HP_*(\mathbb{C}[\Gamma]) \cong H_*(\Gamma, \mathbb{C}).$$

However, the local cyclic homology of $C_{\mathrm{red}}^*\Gamma$ is still unknown.

A Locally convex algebras

A locally convex algebra A is, in general, an algebra with a locally convex topology for which the multiplication $A \times A \to A$ is (jointly) continuous. In the present survey we restrict our attention to locally convex algebras that can be represented as projective limits of Banach algebras.

A locally convex algebra $\mathcal{A}$ that can be represented as a projective limit of Banach algebras can equivalently be defined as a complete locally convex space whose topology is determined by a family $\{p_\alpha\}$ of submultiplicative seminorms, [37]. Thus for each α we have $p_\alpha(xy) \leq p_\alpha(x) p_\alpha(y)$. The algebra $\mathcal{A}$ is then automatically a topological algebra, i.e., multiplication is (jointly) continuous. **We call such algebras m-algebras.**

The unitization $\tilde{\mathcal{A}}$ of an m-algebra is again an m-algebra with the seminorms $\tilde{p}$ given by $\tilde{p}(\lambda 1 + x) = |\lambda| + p(x)$ for all submultiplicative seminorms p on $\mathcal{A}$.

The direct sum $\mathcal{A} \oplus \mathcal{B}$ of two m-algebras is again an m-algebra with the topology defined by all seminorms $p \oplus q$ with $(p \oplus q)(x, y) = p(x) + q(y)$, where p is a continuous seminorm on $\mathcal{A}$ and q a continuous seminorm on $\mathcal{B}$.

Recall the definition of the projective tensor product by Grothendieck, [21, 49]. For two locally convex vector spaces V and W the projective topology on the tensor product $V \otimes W$ is given by the family of seminorms $p \otimes q$, where p is a continuous seminorm on V and q a continuous seminorm on W. Here $p \otimes q$ is defined by

$$p \otimes q(z) = \inf \left\{ \sum_{i=1}^n p(a_i) q(b_i) \;\Big|\; z = \sum_{i=1}^n a_i \otimes b_i, a_i \in V, b_i \in W \right\}$$

for $z \in V \otimes W$. We denote by $V \hat{\otimes} W$ the completion of $V \otimes W$ with respect to this family of seminorms. For m-algebras $\mathcal{A}$ and $\mathcal{B}$, the projective tensor product $\mathcal{A} \hat{\otimes} \mathcal{B}$ is again an m-algebra (if p and q are submultiplicative, so is $p \otimes q$). Also the quotient (completed if necessary) of an m-algebra by a closed ideal is again an m-algebra.

We now list some important examples of m-algebras and of constructions with m-algebras that will be used later on.

A.1 Algebras of differentiable functions

Let $[a,b]$ be an interval in $\mathbb{R}$. We denote by $\mathbb{C}[a,b]$ the algebra of complex-valued $\mathcal{C}^\infty$-functions f on $[a,b]$, all of whose derivatives vanish in a and in b (while f itself may take arbitrary values in a and b).

Also the subalgebras $\mathbb{C}(a,b]$, $\mathbb{C}[a,b)$ and $\mathbb{C}(a,b)$ of $\mathbb{C}[a,b]$, which, by definition consist of functions f, that vanish in a, in b, or in a and b, respectively, will play an important role.

The topology on these algebras is the usual Fréchet topology, which is defined by the following family of submultiplicative norms p_n:

$$p_n(f) = \|f\| + \|f'\| + \tfrac{1}{2}\|f''\| + \cdots + \tfrac{1}{n!}\|f^{(n)}\|$$

Here of course $\|g\| = \sup\{|g(t)| \,|\, t \in [a,b]\}$.

We note that $\mathbb{C}[a,b]$ is nuclear in the sense of Grothendieck [21] and that, for any complete locally convex space V, the space $\mathbb{C}[a,b]\hat{\otimes}V$ is isomorphic to the space of $\mathcal{C}^\infty$-functions on $[a,b]$ with values in V, whose derivatives vanish in both endpoints, [49], § 51. Exactly the same comments apply to the algebra $\mathcal{C}^\infty(M)$ of smooth functions on a compact smooth manifold where an m-algebra topology is defined locally by seminorms like the p_n above.

Given an m-algebra $\mathcal{A}$, we write $\mathcal{A}[a,b]$, $\mathcal{A}[a,b)$ and $\mathcal{A}(a,b)$ for the m-algebras $\mathcal{A}\hat{\otimes}\mathbb{C}[a,b]$, $\mathcal{A}\hat{\otimes}\mathbb{C}[a,b)$ and $\mathcal{A}\hat{\otimes}\mathbb{C}(a,b)$ (their elements are $\mathcal{A}$-valued $\mathcal{C}^\infty$—functions whose derivatives vanish at the endpoints).

Two continuous homomorphisms $\alpha, \beta : A \to B$ between complete locally convex algebras are called differentiably homotopic, or *diffotopic*, if there is a family $\varphi_t : A \to B$, $t \in [0,1]$, of continuous homomorphisms, such that $\varphi_0 = \alpha$, $\varphi_1 = \beta$ and such that the map $t \mapsto \varphi_t(x)$ is infinitely often differentiable for each $x \in A$. An equivalent condition is that there should be a continuous homomorphism $\varphi : A \to \mathcal{C}^\infty([0,1])\hat{\otimes}B$ such that $\varphi(x)(0) = \alpha(x), \varphi(x)(1) = \beta(x)$ for each $x \in A$.

Let $h : [0,1] \to [0,1]$ be a monotone and bijective $\mathcal{C}^\infty$-map, whose restriction to $(0,1)$ gives a diffeomorphism $(0,1) \to (0,1)$ and whose derivatives in 0 and 1 all vanish. Replacing φ_t by $\psi_t = \varphi_{h(t)}$ one sees that α and β are diffotopic if and only if there is a continuous homomorphism $\psi : A \to \mathbb{C}[0,1]\hat{\otimes}B$ such that $\psi(x)(0) = \alpha(x), \psi(x)(1) = \beta(x)$, $x \in A$. This shows in particular that diffotopy is an equivalence relation.

Note that this definition of differentiable homotopy applies not only to m-algebras but to arbitrary complete locally convex algebras. In particular, it applies to any algebra A over $\mathbb{C}$ if we regard it as a locally convex space with the topology given by all possible semi-norms on A. In this case $\mathbb{C}[0,1]\hat{\otimes}A$ is simply the algebraic tensor product $\mathbb{C}[0,1] \otimes A$.

A.2 The smooth tensor algebra

Let V be a complete locally convex space. We define the smooth tensor algebra T^sV as the completion of the algebraic tensor algebra (see Sect. 2.1)

$$TV = V \oplus V {\otimes} V \oplus V^{\otimes^3} \oplus \ldots$$

with respect to the family $\{\widehat{p}\}$ of seminorms, which are given on this direct sum as

$$\widehat{p} = p \oplus p{\otimes}p \oplus p^{\otimes^3} \oplus \ldots$$

where p runs through all continuous seminorms on V. The seminorms $\widehat{p}$ are submultiplicative for the multiplication on TV. The completion T^sV therefore is an m-algebra.

In the simplest case $V = \mathbb{C}$, $T^s\mathbb{C}$ is naturally isomorphic to the algebra of holomorphic functions on the complex plane that vanish at 0 (under the isomorphism which maps a sequence (λ_n) in $T\mathbb{C}$ to the function f with $f(z) = \sum_{n=1}^{\infty} \lambda_n z^n$). The topology is the topology of uniform convergence on compact subsets.

We denote by $\sigma : V \to T^sV$ the map, which maps V to the first summand in TV. The map σ has the following universal property: let $s : V \to \mathcal{A}$ be any continuous linear map from V to an m-algebra $\mathcal{A}$. Then there is a unique continuous homomorphism $\tau_s : T^sV \to \mathcal{A}$ of m-algebras such that $\tau_s \circ \sigma = s$. It is given by

$$\tau_s(x_1 \otimes x_2 \otimes \ldots \otimes x_n) = s(x_1)s(x_2) \ldots s(x_n)$$

The smooth tensor algebra is differentiably contractible, i.e., the identity map on T^sV is diffotopic to 0. A differentiable family $\varphi_t : T^sV \to T^sV$ of homomorphisms for which $\varphi_0 = 0, \varphi_1 = \mathrm{id}$ is given by $\varphi_t = \tau_{t\sigma}, t \in [0,1]$.

If $\mathcal{A}$ is an m-algebra, by abuse of notation, we write $T\mathcal{A}$ (rather than $T^s\mathcal{A}$) for the smooth tensor algebra over $\mathcal{A}$. Thus $T\mathcal{A}$ is again an m-algebra.

A.3 The free product of two m-algebras

Let $\mathcal{A}$ and $\mathcal{B}$ be m-algebras. The (non-unital) algebraic free product $\mathcal{A} * \mathcal{B}$ is the direct sum over all tensor products

$$\mathcal{A} * \mathcal{B} = \mathcal{A} \oplus \mathcal{B} \oplus (\mathcal{A}{\otimes}\mathcal{B}) \oplus (\mathcal{B}{\otimes}\mathcal{A}) \oplus (\mathcal{A}{\otimes}\mathcal{B}{\otimes}\mathcal{A}) \oplus \ldots$$

where the factors $\mathcal{A}$ and $\mathcal{B}$ alternate (see 2.1). Multiplication is, as in 2.1, concatenation of tensors combined with simplification using the multiplication in $\mathcal{A}$ and $\mathcal{B}$.

We denote by $\mathcal{A}\hat{*}\mathcal{B}$ the completion of $\mathcal{A} * \mathcal{B}$ with respect to the family of all seminorms of the form $p * q$, given by:

$$p * q = p \oplus q \oplus (p \otimes q) \oplus (q \otimes p) \oplus (p \otimes q \otimes p) \oplus \ldots$$

where p and q are continuous seminorms on $\mathcal{A}$ and $\mathcal{B}$, respectively. If p and q are submultiplivative, then so is the seminorm $p * q$ and $\mathcal{A}\hat{*}\mathcal{B}$ therefore is an m-algebra.

$\mathcal{A}\hat{*}\mathcal{B}$ is in fact the free product of $\mathcal{A}$ and $\mathcal{B}$ in the category of m-algebras. The canonical inclusions $\iota_1 : \mathcal{A} \to \mathcal{A}\hat{*}\mathcal{B}$ and $\iota_2 : \mathcal{B} \to \mathcal{A}\hat{*}\mathcal{B}$ have the following universal property: Let $\alpha : \mathcal{A} \to \mathcal{E}$ and $\beta : \mathcal{B} \to \mathcal{E}$ be two continuous homomorphisms into an m-algebra $\mathcal{E}$. Then there is a unique continuous homomorphism $\alpha * \beta : \mathcal{A}\hat{*}\mathcal{B} \to \mathcal{E}$, such that $(\alpha * \beta) \circ \iota_1 = \alpha$ and $(\alpha * \beta) \circ \iota_2 = \beta$.

If $\mathcal{A}$ and $\mathcal{B}$ are m-algebras, by abuse of notation, we write $\mathcal{A} * \mathcal{B}$ (rather than $\mathcal{A}\hat{*}\mathcal{B}$) for the m-algebra free product of $\mathcal{A}$ and $\mathcal{B}$.

A.4 The algebra of smooth compact operators

The m-algebra $\mathcal{K}$ of "smooth compact operators" consists of all $\mathbb{N} \times \mathbb{N}$-matrices (a_{ij}) with rapidly decreasing matrix elements $a_{ij} \in \mathbb{C}$, $i,j = 0,1,2\ldots$. The topology on $\mathcal{K}$ is given by the family of norms p_n, $n = 0,1,2\ldots$, which are defined by

$$p_n((a_{ij})) = \sum_{i,j} |1+i+j|^n |a_{ij}|$$

It is easily checked that the p_n are submultiplicative and that $\mathcal{K}$ is complete. Thus $\mathcal{K}$ is an m-algebra. As a locally convex vector space, $\mathcal{K}$ is isomorphic to the space s of rapidly decreasing functions and therefore nuclear.

The map that sends $(a_{ij}) \otimes (b_{kl})$ to the $\mathbb{N}^2 \times \mathbb{N}^2$-matrix $(a_{ij}b_{kl})_{(i,k)(j,l) \in \mathbb{N}^2 \times \mathbb{N}^2}$ obviously gives an isomorphism Θ between $\mathcal{K}\hat{\otimes}\mathcal{K}$ and $\mathcal{K}$ (see also [41], 2.7).

Lemma A.1. *Let $\Theta : \mathcal{K} \to \mathcal{K}\hat{\otimes}\mathcal{K}$ be as above and let $\iota : \mathcal{K} \to \mathcal{K}\hat{\otimes}\mathcal{K}$ be the inclusion that maps x to $e^{00} \otimes x$ (where e^{00} is the matrix with elements a_{ij}, for which $a_{ij} = 1$, whenever $i = j = 0$, and $a_{ij} = 0$ otherwise). Then Θ is diffotopic to ι. The same holds for the corresponding maps $\Theta' : \mathcal{K} \to M_2(\mathcal{K})$ and $\iota' : \mathcal{K} \to M_2(\mathcal{K})$.*

The proof uses an identification of $\mathcal{K}$ with $\mathcal{S}(\mathbb{R})\hat{\otimes}\mathcal{S}(\mathbb{R}) \cong \mathbb{C}(0,1)\hat{\otimes}\mathbb{C}(0,1)$.

Remark A.2. Let V be a Banach space. Then $\mathcal{K}\hat{\otimes}V$ consists exactly of those matrices (or the sequences indexed by $\mathbb{N} \times \mathbb{N}$) $(v_{ij})_{i,j \in \mathbb{N}}$, for which the expression

$$\bar{p}_n((v_{ij})) \underset{def}{=} \sum_{i,j} (1+i+j)^n \|v_{ij}\|$$

is finite for all n. The topology on $\mathcal{K}\hat{\otimes}V$ is of course given exactly by the norms $\bar{p}_n$. To see this, consider the tensor product α_n of the norm p_n on $\mathcal{K}$ with the norm $\|\cdot\|$ given on V. Then, if x^{ij} denotes the matrix, which has $x \in V$ as element on the place i,j and is 0 otherwise, we have

$$\alpha_n(x^{ij}) = (1+i+j)\|x\|$$

by [49], Prop. 43.1. This immediately shows that

$$\alpha_n((v_{ij})) \leq \bar{p}_n((v_{ij}))$$

for all matrices (v_{ij}) in the algebraic tensor product $\mathcal{K} \otimes V$. The converse inequality follows from the definition of the projective tensor product. Therefore, for each fixed n, the completion $(\mathcal{K} \otimes V)_{\bar{p}_n}$ is isometrically isomorphic to $(\mathcal{K})_{p_n} \hat{\otimes} V$ and consists exactly of those matrices (v_{ij}), for which $\bar{p}_n((v_{ij}))$ is finite.

A.5 The Schatten ideals $\ell^p(H)$

Let H be an infinite-dimensional separable Hilbert space (i.e., $H \cong \ell^2(\mathbb{N})$), and $(\xi_n)_{n \in \mathbb{N}}$ an orthonormal basis. Let also $\mathcal{L}(H)$ denote the algebra of bounded operators on H. For each positive x in $\mathcal{L}(H)$, we can define the trace $\operatorname{Tr}(x)$ by
$$\operatorname{Tr}(x) = \sum_n (x\xi_n \mid \xi_n)$$
This expression is ≥ 0 and may be infinite. It does not depend on the choice of the orthonormal basis. Moreover $\operatorname{Tr}(x^*x) = \operatorname{Tr}(xx^*)$ holds for all x.

Definition A.1 *Let $p \geq 1$. We define $\ell^p(H)$ to be the set of all operators $x \in \mathcal{L}(H)$ for which*
$$\operatorname{Tr} |x|^p < \infty$$
where $|x| = \sqrt{x^*x}$.

Proposition A.2 *For each $p \geq 1$, $\ell^p(H)$ is an ideal in $\mathcal{L}(H)$ which is contained in the ideal of compact operators on H. It is a Banach algebra with respect to the norm*
$$\|x\|_p = (\operatorname{Tr} |x|^p)^{1/p}$$
Moreover $\ell^p(H) \subset \ell^q(H)$ for $p \leq q$

A.6 The smooth Toeplitz algebra

The elements of the algebra $C^\infty S^1$ can be written as power series in the generator z (defined by $z(t) = t$, $t \in S^1 \subset \mathbb{C}$). The coefficients of these power series are rapidly decreasing. Thus
$$C^\infty(S^1) = \left\{ \sum_{k \in \mathbb{Z}} a_k z^k \ \Big| \ \sum_{k \in \mathbb{Z}} |a_k| |k|^n < \infty \text{ for all } n \in \mathbb{N} \right\}$$

Submultiplicative norms that determine the topology are given by
$$q_n \left(\sum a_k z^k \right) = \sum |1 + k|^n |a_k|$$

As a topological vector space, the smooth Toeplitz algebra $\mathcal{T}$ may be defined as the direct sum $\mathcal{T} = \mathcal{K} \oplus C^\infty(S^1)$.

To describe the multiplication in $\mathcal{T}$, we write v_k for the element $(0, z^k)$ in $\mathcal{T}$ and simply x for the element $(x, 0)$ with $x \in \mathcal{K}$. Moreover, e^{ij} denotes

the element of $\mathcal{T}$, which is determined by the matrix (a_{kl}), where $a_{kl} = 1$, whenever $k = i, l = j$, and $a_{kl} = 0$ otherwise (with the convention that $e^{ij} = 0$, if $i < 0$ oder $j < 0$). Multiplication in $\mathcal{T}$ then is determined by the following rules:
$$e^{ij}e^{kl} = \delta_{jk}e^{il}$$
$$v_k e^{ij} = e^{(i+k),j} \qquad e^{ij}v_k = e^{i,(j-k)}$$
$(i, j, k \in \mathbb{Z})$; and
$$v_k v_{-l} = \begin{cases} v_{k-l}(1 - E_{l-1}) & l > 0 \\ v_{k-l} & l \leq 0 \end{cases}$$

where $E_l = e^{00} + e^{11} + \cdots + e^{ll}$. If p_n are the norms on $\mathcal{K}$ defined in A.4 and q_n the norms on $\mathcal{C}^\infty(S^1)$ defined above, it is easy to see that each norm $p_n \oplus q_n$ is submultiplicative on $\mathcal{T} = \mathcal{K} \oplus \mathcal{C}^\infty(S^1)$ with the multiplication introduced above. Obviously, $\mathcal{K}$ is a closed ideal in $\mathcal{T}$ and the quotient $\mathcal{T}/\mathcal{K}$ is $\mathcal{C}^\infty(S^1)$.

B Standard extensions

Extensions play a fundamental role in all aspects of non-commutative geometry. In this section we describe standard extensions that can be associated to any given algebra. All of these extensions exist in a purely algebraic setting, but also in the category of m-algebras. We explicitly treat here only the case of m-algebras.

B.1 The suspension extension

This is the fundamental extension of algebraic topology. It has the form
$$0 \to \mathbb{C}(0,1) \to \mathbb{C}[0,1) \to \mathbb{C} \to 0$$
or more generally
$$0 \to \mathcal{A}(0,1) \to \mathcal{A}[0,1) \to \mathcal{A} \to 0$$
with an arbitrary m-algebra $\mathcal{A}$.

Since cyclic homology is homotopy invariant only for differentiable homotopies (not for continuous ones) we have to work with algebras of differentiable functions. Recall from A.1 that $\mathbb{C}(0,1)$ and $\mathbb{C}[0,1)$ denote the algebras of $\mathcal{C}^\infty$—functions on the interval $[0,1]$, whose derivatives to arbitrary order vanish in 0 and 1 and that the algebra $\mathbb{C}[0,1)$ is differentiably contractible.

B.2 The free extension

For any m-algebra $\mathcal{A}$ there exists a natural extension

$$0 \to J\mathcal{A} \to T\mathcal{A} \xrightarrow{\pi} \mathcal{A} \to 0. \qquad (1)$$

by the completion $T\mathcal{A}$ of the free algebra $\mathcal{A} \oplus \mathcal{A}^{\otimes 2} \oplus \ldots$ — the nonunital smooth tensor algebra over $\mathcal{A}$, cf. A.2. Here π maps a tensor $x_1 \otimes x_2 \otimes \ldots \otimes x_n$ to $x_1 x_2 \ldots x_n \in \mathcal{A}$ and $J\mathcal{A} = \mathrm{Ker}\,\pi$. This extension is universal in the sense that, given any extension $0 \to \mathcal{I} \to \mathcal{E} \to \mathcal{A} \to 0$ of A, admitting a continuous linear splitting, there is a morphism of extensions

$$\begin{array}{ccccccccc} 0 \to & J\mathcal{A} & \to & T\mathcal{A} & \to & \mathcal{A} & \to 0 \\ & \downarrow \gamma & & \downarrow \tau & & \downarrow id & \\ 0 \to & \mathcal{I} & \to & \mathcal{E} & \to & \mathcal{A} & \to 0 \end{array}$$

The map $\tau : T\mathcal{A} \to \mathcal{E}$ is obtained by choosing a continuous linear splitting $s : \mathcal{A} \to \mathcal{E}$ in the given extension and mapping $x_1 \otimes x_2 \otimes \ldots \otimes x_n$ to $s(x_1)s(x_2)\ldots s(x_n) \in E$. Then γ is the restriction of τ.

Definition B.1. *The map* $\gamma : J\mathcal{A} \to \mathcal{I}$ *in this commutative diagram is called the* classifying map *for the extension* $0 \to \mathcal{I} \to \mathcal{E} \to \mathcal{A} \to 0$.

If s and s' are two different linear sections, then for each $t \in [0,1]$ the map $s_t = ts + (1-t)s'$ is again a splitting. The corresponding maps γ_t associated with s_t form a homotopy between the classifying map constructed from s and the one constructed from s'. The classifying map is therefore unique up to homotopy.

More generally, any n-step extension of the form

$$0 \to \mathcal{I} \to \mathcal{E}_1 \to \mathcal{E}_2 \to \ldots \to \mathcal{E}_n \to \mathcal{A} \to 0$$

admitting a continuous linear splitting can be compared with the free n-step extension

$$0 \to J^n \mathcal{A} \to T(J^{n-1}\mathcal{A}) \to T(J^{n-2}\mathcal{A}) \to \ldots \to T\mathcal{A} \to \mathcal{A} \to 0$$

and thus has a classifying map $J^n \mathcal{A} \to \mathcal{I}$ which is unique up to homotopy (here $J^n \mathcal{A}$ is defined recursively by $J^n \mathcal{A} = J(J^{n-1}\mathcal{A})$).

B.3 The universal two-fold trivial extension.

With any m-algebra $\mathcal{A}$ we associate as in [11] the m-algebra $Q\mathcal{A} = \mathcal{A} * \mathcal{A}$. We denote by ι and $\bar{\iota}$ the two canonical inclusions of $\mathcal{A}$ into $Q\mathcal{A}$. The algebra $Q\mathcal{A}$ is, in a natural way, $\mathbb{Z}/2$-graded by the involutive automorphism τ which exchanges $\iota(\mathcal{A})$ and $\bar{\iota}(\mathcal{A})$.

The ideal $q\mathcal{A}$ in $Q\mathcal{A}$ is by definition the kernel of the homomorphism $\pi = \mathrm{id} * \mathrm{id} : \mathcal{A} * \mathcal{A} \to \mathcal{A}$. The extension

$$0 \to q\mathcal{A} \to Q\mathcal{A} \xrightarrow{\pi} \mathcal{A} \to 0 \qquad (2)$$

admits two natural splittings via the algebra homomorphisms ι and $\bar{\iota}$. It has the following universal property: Let

$$0 \to \mathcal{E}_0 \to \mathcal{E}_1 \to \mathcal{A} \to 0 \qquad (3)$$

be an extension with two splittings $\alpha, \bar{\alpha} : \mathcal{A} \to \mathcal{E}_1$, which are continuous algebra homomorphisms. Then there is a morphism of extensions (i.e., a commutative diagram) as follows:

$$\begin{array}{ccccccccc} 0 \to & q\mathcal{A} & \to & Q\mathcal{A} & \to & \mathcal{A} & \to & 0 \\ & \downarrow \alpha*\bar{\alpha} & & \downarrow \alpha*\bar{\alpha} & & \downarrow \mathrm{id} & & \\ 0 \to & \mathcal{E}_0 & \to & \mathcal{E}_1 & \to & \mathcal{A} & \to & 0 \end{array}$$

By construction, this morphism sends ι and $\bar{\iota}$ to α and $\bar{\alpha}$.

The same construction can of course be performed purely algebraically and we obtain, for any algebra A, an extension

$$0 \to qA \to QA \to A \to 0$$

where $QA = A * A$. This extension has the same universal property in the category of algebras without topology.

B.4 The Toeplitz extension

The smooth Toeplitz algebra $\mathcal{T}$ was introduced in A.6. By definition, $\mathcal{T}$ contains the algebra $\mathcal{K}$ of smooth compact operators as a closed ideal and we obtain the following extension (Toeplitz extension)

$$0 \to \mathcal{K} \to \mathcal{T} \xrightarrow{\pi} \mathcal{C}^\infty(S^1) \to 0$$

which, of course, admits a continuous linear splitting.

We consider now $\mathbb{C}(0,1)$ as a subalgebra of $\mathcal{C}^\infty(S^1)$ and denote by $\mathcal{T}_0$ the preimage of $\mathbb{C}(0,1)$ under π. By restriction we obtain the following extension

$$0 \to \mathcal{K} \to \mathcal{T}_0 \to \mathbb{C}(0,1) \to 0$$

Taking the Yoneda product with the suspension extension

$$0 \longrightarrow \mathbb{C}(0,1) \longrightarrow \mathbb{C}[0,1] \longrightarrow \mathbb{C} \longrightarrow 0$$

gives the following extension of length 2

$$0 \to \mathcal{K} \to \mathcal{T}_0 \to \mathbb{C}[0,1] \to \mathbb{C} \to 0$$

More generally we can tensor this extension with an arbitrary m-algebra $\mathcal{A}$ and obtain the following standard two step extension (Bott extension) for $\mathcal{A}$

$$0 \to \mathcal{K}\hat{\otimes}\mathcal{A} \to \mathcal{T}_0\hat{\otimes}\mathcal{A} \to \mathcal{A}[0,1) \to \mathcal{A} \to 0$$

References

1. B. Blackadar: K-theory for Operator Algebras, second Edition, MSRI Publications, Cambridge University Press, 1998.
2. B. Blackadar and J. Cuntz. Differential Banach algebra norms and smooth subalgebras of C^*-algebras. *J. Operator Theory*, **26**(2) (1991), 255–282.
3. L.G. Brown, R.G. Douglas und P. Fillmore: Extensions of C*-algebras and K-homology, Ann. of Math. **105** (1977), 265–324.
4. A. Connes: Non-commutative differential geometry. Publ. Math. I.H.E.S. **62** (1985), 257–360.
5. A. Connes: Entire cyclic cohomology of Banach algebras and characters of θ-summable Fredholm modules. *K-Theory*, **1**(6) (1998), 519–548.
6. A. Connes. On the Chern character of θ summable Fredholm modules. *Comm. Math. Phys.*, **139**(1) (1991), 171–181.
7. A. Connes, Non-commutative Geometry, Academic Press, London-Sydney-Tokyo-Toronto, 1994.
8. A. Connes and J. Cuntz: Quasi-homomorphismes, cohomologie cyclique et positivité, Commun. Math. Phys. **114** (1988), 515–526.
9. A. Connes and N. Higson: Déformations, morphismes asymptotiques et K-théorie bivariante, C.R.Acad.Sci. Paris Ser. I, Math, **311** (1990), 101–106.
10. G. Cortinas and C. Valqui: Excision in bivariant periodic cyclic cohomology: A categorical approach, preprint 2002.
11. J. Cuntz: A new look at KK-theory, K-theory **1** (1987), 31-52.
12. J. Cuntz: Excision in bivariant cyclic theory for topological algebras, Fields Institute Communications **vol 17** 1997 (edts. J. Cuntz-M. Khalkhali) AMS, 43-52.
13. J. Cuntz: Bivariante K-theorie für lokalkonvexe Algebren und der bivariante Chern-Connes-Charakter, Docum. Math. J. DMV **2** (1997), 139–182, http://www.mathematik.uni-bielefeld.de/documenta
14. J. Cuntz: Morita invariance in cyclic homology for nonunital algebras, K-theory **15** (1998), 301–305.
15. J. Cuntz and D. Quillen, Algebra extensions and nonsingularity, J.Amer. Math. Soc. **8** (1995), 251–289.
16. J. Cuntz and D. Quillen: Cyclic homology and nonsingularity, J. Amer. Math. Soc. **8** (1995), 373–442.
17. J. Cuntz and D. Quillen: Excision in bivariant periodic cyclic cohomology, Invent. math. **127** (1997), 67–98.
18. J. Cuntz and D. Quillen, Operators on noncommutative differential forms and cyclic homology, in "Geometry, Topology and Physics; for Raoul Bott", International Press, Cambridge MA, 1995.
19. G.A. Elliott, T. Natsume and R. Nest: Cyclic cohomology for one-parameter smooth crossed products, Acta Math. **160** (1988), 285–305.
20. T.G. Goodwillie: Cyclic homology and the free loopspace. Topology **24** (1985), 187–215.
21. A. Grothendieck: Produits tensoriels topologiques et espaces nucléaires, Memoirs of the Amer. Math. Soc. **16** (1955).
22. J.A. Guccione and J.J. Guccione: The theorem of excision for Hochschild and cyclic homology, J. Pure Appl. Algebra **106** (1996), 57–60.
23. N. Higson: On Kasparov theory, M.A. thesis, Dalhousie University, 1983.

24. H. Hogbe-Nlend: *Bornologies and functional analysis*: North-Holland Publishing Co., Amsterdam, 1977.
25. A. Jaffe, A. Lesniewski, and K. Osterwalder: Quantum K-theory. I. The Chern character. *Comm. Math. Phys.*, **118**(1) (1988), 1–14.
26. J.D.S. Jones and C. Kassel: Bivariant cyclic theory, K-theory **3** (1989), 339–365.
27. M. Karoubi, Homologie cyclique et K-théorie, Astérisque **149** (1987).
28. G.G. Kasparov, The operator K-functor and extensions of C^*-algebras, Izv. Akad. Nauk. SSSR, Ser. Mat. **44** (1980), 571–636.
29. C. Kassel, Cyclic homology, comodules and mixed complexes, J. of Algebra **107** (1987), 195–216.
30. C. Kassel, Caractère de Chern bivariant. K-Theory **3** (1989), 367–400.
31. M. Khalkhali. Algebraic connections, universal bimodules and entire cyclic cohomology. *Comm. Math. Phys.*, **161**(3) (1994), 433–446.
32. V. Lafforgue: K-théorie bivariante pour les algèbres de Banach et conjecture de Baum-Connes, thesis, Paris 1999.
33. J.-L. Loday, Cyclic Homology, Grundlehren **301**, Springer Verlag, 1992.
34. J.-L. Loday and D. Quillen, Cyclic homology and the Lie algebra homology of matrices, Comment. Math. Helvetici **59** (1984), 565–591.
35. R. Meyer. Analytic cyclic cohomology. Thesis. math.KT/9906205, 1999.
36. R. Meyer, Excision in entire cyclic cohomology, J. Eur. Math. Soc. **3** (2001), 269–286.
37. E.A. Michael, Locally multiplicatively-convex topological algebras, Memoirs of the AMS, number **11**, 1952.
38. V. Nistor, A bivariant Chern character for p-summable quasihomomorphisms. K-Theory **5** (1991), 193–211.
39. V. Nistor, A bivariant Chern-Connes character, Ann. of Math. **138** (1993), 555–590.
40. D. Perrot: A bivariant Chern character for families of spectral triples, Preprint math-ph/0103016, 2001.
41. C. Phillips: K-theory for Fréchet algebras, International Journal of Mathematics vol. **2**(1) (1991), 77–129.
42. M. Puschnigg: Asymptotic cyclic cohomology, Lecture Notes in Mathematics **1642**, Springer 1996.
43. M. Puschnigg: Cyclic homology theories for topological algebras, preprint.
44. M. Puschnigg: Excision in cyclic homology theories, Invent. Math. **143** (2001), 249–323.
45. M. Puschnigg: The Kadison-Kaplansky conjecture for word-hyperbolic groups, Invent. Math. **149** (2002), 153–194.
46. D. Quillen, Algebra cochains and cyclic cohomology, Publ. Math. IHES, **68** (1988), 139–174.
47. E. Schrohe, M. Walze, and J.-M. Warzecha. Construction de triplets spectraux à partir de modules de Fredholm. *C. R. Acad. Sci. Paris Sér. I Math.*, **326**(10) (1998), 1195–1199.
48. W.F. Schelter: Smooth algebras, J. Algebra **103**(1) (1986), 677–685.
49. F. Treves: Topological vector spaces, distributions and kernels, Academic Press, New York, London, 1967.
50. B. Tsygan: The homology of matrix Lie algebras over rings and the Hochschild homology (in Russian), Uspekhi Mat. Nauk **38** (1983), 217–218; Russ. Math. Survey **38**(2) (1983), 198–199.

51. J.-L. Tu. La conjecture de Baum-Connes pour les feuilletages moyennables. K-Theory, **17**(3) (1999), 215–264.
52. X. Wang, A bivariant Chern character, II, Can. J. Math. **35** (1992), 1–36.
53. M. Wodzicki, The long exact sequence in cyclic homology associated with an extension of algebras, C.R. Acad. Sci. Paris **306** (1988), 399–403.
54. M. Wodzicki, Excision in cyclic homology and in rational algebraic K-theory, Ann. of Math. **129** (1989), 591-639.

Cyclic Homology*

Boris Tsygan

1	**Introduction**	74
2	**Hochschild and cyclic homology of algebras**	76
2.1	Homology of differential graded algebras	78
2.2	The Hochschild cochain complex	79
2.3	Products on Hochschild and cyclic complexes	80
2.4	Pairings between chains and cochains	82
2.5	A_∞ structure on $C_\bullet(C^\bullet(A))$	84
3	**The cyclic complex $C_\bullet^\lambda$**	86
3.1	Definition	86
3.2	The reduced cyclic complex	87
3.3	Relation to Lie algebra homology	88
3.4	The connecting morphism	88
3.5	Other operations on the cyclic complexes	90
3.6	Rigidity of periodic cyclic homology	91
4	**Noncommutative differential calculus**	92
4.1	Gerstenhaber algebras	92
4.2	The Gerstenhaber algebra $\mathcal{V}^\bullet(A)$	92
4.3	Calculi	93
4.4	The calculus $\mathrm{Calc}(A)$	94
4.5	Enveloping algebra of a Gerstenhaber algebra	95
4.6	Relation to the A_∞ structure on chains	95
5	**Cyclic objects**	96
5.1	Simplicial objects	96
5.2	Cyclic objects	96

* Supported in part by NSF grant.

6	**Examples**	98
6.1	Smooth functions	98
6.2	Commutative algebras	99
6.3	Rings of differential operators	100
6.4	Rings of complete symbols	101
6.5	Rings of pseudodifferential operators	101
6.6	Noncommutative tori	102
6.7	Deformation quantization	102
6.8	The Brylinski spectral sequence	105
6.9	Deformation quantization: general case	105
6.10	Group rings	106
7	**Index theorems**	107
7.1	Index theorem for deformations of symplectic structures	107
7.2	Index theorem for holomorphic symplectic deformations	108
7.3	General index theorem for deformations	108
8	**Riemann-Roch theorem for D-modules**	109
References		110

1 Introduction

Many geometric objects associated to a manifold M can be expressed in terms of an appropriate algebra A of functions on M (measurable, continuous, smooth, holomorphic, algebraic, ...). Very often those objects can be defined in a way that is applicable to any algebra A, commutative or not. Study of associative algebras by means of such objects of geometric origin is the subject of noncommutative geometry [12, 48]. The Hochschild and cyclic (co)homology theory is the part of noncommutative geometry which generalizes the classical differential and integral calculus. The geometric objects being generalized to the noncommutative setting are differential forms, densities, multivector fields, etc.

In our exposition, the primary object is the negative cyclic complex $CC_\bullet^-(A)$. Other complexes, namely the Hochschild chain complex $C_\bullet(A)$, the periodic cyclic complex $CC_\bullet^{\text{per}}(A)$, and the cyclic complex $CC_\bullet(A)$, are defined as results of some natural procedure applied to $CC_\bullet^-(A)$. The cyclic homology is the homology of the cyclic complex $CC_\bullet(A)$. It was originally defined using another standard complex which we denote by $C_\bullet^\lambda(A)$. The study of this latter complex has a distinctly different flavor, mainly coming from the fact that it is related to the Lie algebra homology.

The above complexes are noncommutative versions of the space of differential forms (the Hochschild chain complex) and of the De Rham complex. One also defines the Hochschild cochain complex $C^\bullet(A, A)$ which is a noncommutative analogue of the space of multivector fields.

In the first three sections following this Introduction, we study rather systematically various algebraic structures on the above complexes. These structures are supposed to generalize the classical algebraic structures arising in calculus. Sections 2 and 4 are devoted to the structures on the complex $CC_\bullet^-(A)$ and related complexes; Sect. 3 deals with the cyclic complex $C_\bullet^\lambda(A)$.

These algebraic structures can be studied on three levels of complexity. One can start by defining basic algebaric operations generalizing the product and the bracket of multivector fields and various pairings between multivectors and forms; these are operations at the level of complexes. They acquire the standard prorerties from calculus only if one passes from complexes to their cohomology. When the algebra A is the algebra of functions on a manifold, then the homology of the Hochschild chain complex is the space of forms, the cohomology of the Hochschild cochain complex is the space of multivectors, etc. So, in a more naive sense, even at the first level of noncommutative calculus we already achieve our goal: to generalize the basic definitions and structures of calculus to the noncommutative setting (Sects. 2.3, 2.4). But this generalization is not interesting when the algebra becomes noncommutative. For example, for the algebra of differential operators on $\mathbb{R}^n$ both the spaces of "noncommutative forms" and "noncommutative multivectors" are one-dimensional.

To make noncommutative calculus useful for applications, one has to pass to the second level of complexity. Namely, one tries to reproduce more of the standard algebraic structure from the classical calculus not at the level of cohomology but rather at the level of complexes. This plan succeeds partially, in the following sense: one can define certain algebraic structures on Hochschild chains and cochains. These structures are only part of what one could expect from the classical calculus. The advantage of these structures is that they are defined by explicit, canonical constructions (2.5, 3.5). They are also adequate for some applications, namely for the index theorems for symplectic deformations.

To construct a richer algebraic structure on Hochschild chains and cochains, one that generalizes a fuller version of the classical calculus, one has to go to a different level of complexity. The work in this direction was started by Tamarkin in [64] and then continued in [67–69]. We discuss these methods in Sect. 4. They provide a considerable refinement of the results of 2.5, 3.5, but there is no canonical and explicit construction anymore. The "noncomutative differential calculus" can be constructed using some inexplicit formulas; a choice of coefficients in these formulas depends on a choice of a Drinfeld associator [22]. In particular, the Grothendieck-Teichmüller group acts on the space of all such calculi. This version of noncommutative calculus allows one to generalize the index theorem from symplectic deformations to arbitrary deformations.

Note that in [12, 16–18] a different, though clearly related, version of noncommutative calculus is used. In particular, there the renormalization group appears as a hidden group of symmetries of the calculus, whereas in our construction the Grothendieck-Teichmüller group acts on the space of universal formulas defining a calculus. This group is closely related to the Galois group $\text{Gal}(\overline{\mathbb{Q}}/\mathbb{Q})$. Note that finding a unified symmetry group incorporating both the renormalization group and the Galois group of $\mathbb{Q}$ is one of the important aims of Connes' noncommutative geometry program. In light of this, it seems to be an interesting problem to find a unified framework to the two approaches to noncommutative calculus (the author would like to thank Alain Connes for his remarks on this subject).

In Sect. 6, we study various examples of computations of the Hochschild and cyclic homology. The examples that we choose in this presentation revolve mostly around several related classes of algebras: functions on manifolds or on algebraic varieties; operators on functions; deformed algebras of functions; group algebras. Many interesting examples, in particular the ones related to algebraic topology, are not considered here. We compute the Hochschild homology and various versions of the cyclic homology of any formal deformation of the algebra of smooth functions on a C^∞ manifold. This computation, as well as the classification of all such deformations, is based on the noncommutative differential calculus from Sect. 4. The constructions that we use are neither explicit nor canonical; they depend on a choice of a Drinfeld associator.

In Sect. 7, index theorems for formal deformations are discussed. An index theorem is an equality that compares two maps from the (periodic) cyclic complex of a deformed algebra of functions on a manifold to the De Rham complex of this manifold. One of these maps is the principal symbol; it is, essentially, reduction modulo deformation parameter, followed by the Connes isomorphism between the cyclic homology and the De Rham cohomology. The other is the trace density map. It is given by an explicit and canonical construction in the case when the quasi-classical limit of the deformation is symplectic; in the general case, its definition depends on a choice of an associator.

In Sect. 8 we outline an application of the above theorem (in the symplectic case) to a general index theorem. This theorem, conjectured in [60] and proven in [4], contains many classical index theorems as partial cases, such as the Atiyah-Singer theorem for real analytic elliptic operators, the Boutet de Monvel index theorem for boundary value problems, the Kashiwara index theorem for holonomic D-modules, and others.

2 Hochschild and cyclic homology of algebras

Let k denote a commutative algebra over a field of characteristic zero and let A be a flat k-algebra with unit, not necessarily commutative. Let $\overline{A} = A/k \cdot 1$, and let $C_p(A) \stackrel{def}{=} A \otimes_k \overline{A}^{\otimes_k p}$. Define

$$b : C_p(A) \to C_{p-1}(A) \qquad (1)$$

$$a_0 \otimes \cdots \otimes a_p \mapsto (-1)^p a_p a_0 \otimes \cdots \otimes a_{p-1} +$$

$$\sum_{i=0}^{p-1} (-1)^i a_0 \otimes \cdots \otimes a_i a_{i+1} \otimes \cdots \otimes a_p \ .$$

Then $b^2 = 0$ and one gets the complex $(C_\bullet, b)$, called *the standard Hochschild complex of A*. The homology of this complex is denoted by $H_\bullet(A, A)$, or by $HH_\bullet(A)$.

Proposition 2.1. *The map*

$$B : C_p(A) \to C_{p+1}(A) \qquad (2)$$

$$a_0 \otimes \cdots \otimes a_p \mapsto \sum_{i=0}^{p} (-1)^{pi} 1 \otimes a_i \otimes \cdots \otimes a_p \otimes a_0 \otimes \cdots \otimes a_{i-1}$$

satisfies $B^2 = 0$ and $bB + Bb = 0$ and therefore defines a map of complexes

$$B : C_\bullet(A) \to C_\bullet(A)[-1]$$

Definition 2.2. *For $i, j, p \in \mathbb{Z}$ let*

$$CC_p^-(A) = \prod_{\substack{i \geq p \\ i = p \bmod 2}} C_i(A)$$

$$CC_p^{per}(A) = \prod_{\substack{i = p \bmod 2}} C_i(A)$$

$$CC_p(A) = \bigoplus_{\substack{i \leq p \\ i = p \bmod 2}} C_i(A)$$

The complex $(CC_\bullet^-(A), B + b)$ (respectively $(CC_\bullet^{per}(A), B + b)$, respectively $(CC_\bullet(A), B + b)$) is called the *negative cyclic* (respectively *periodic cyclic*, respectively *cyclic*) complex of A. The homology of these complexes is denoted by $HC_\bullet^-(A)$, respectively by $HC_\bullet^{per}(A)$, respectively by $HC_\bullet(A)$.

There are inclusions of complexes

$$CC_\bullet^-(A)[-2] \hookrightarrow CC_\bullet^-(A) \hookrightarrow CC_\bullet^{per}(A) \qquad (3)$$

and the short exact sequences

$$0 \to CC_\bullet^-(A)[-2] \to CC_\bullet^-(A) \to C_\bullet(A) \to 0 \qquad (4)$$

$$0 \to C_\bullet(A) \to CC_\bullet(A) \xrightarrow{S} CC_\bullet(A)[2] \to 0 \qquad (5)$$

To the double complex $CC_\bullet(A)$ one associates the spectral sequence

$$E^2_{pq} = H_{p-q}(A, A) \tag{6}$$

converging to $HC_{p+q}(A)$.

In what follows we will use the notation of Getzler and Jones ([36]). Let u denote a variable of degree -2. Then the negative and periodic cyclic complexes are described by the following formulas:

$$CC^-_\bullet(A) = (C_\bullet(A)[[u]], b + uB) \tag{7}$$
$$CC^{\text{per}}_\bullet(A) = (C_\bullet(A)[[u, u^{-1}]], b + uB) \tag{8}$$
$$CC_\bullet(A) = (C_\bullet(A)[[u, u^{-1}]]/uC_\bullet(A)[[u]], b + uB) \tag{9}$$

In this language, the map S is just multiplication by u.

Remark 2.3. For an algebra A without unit, let $\tilde{A} = A + k \cdot 1$ and put

$$CC_\bullet(A) = \text{Ker}(CC_\bullet(\tilde{A}) \to CC_\bullet(k));$$

similarly for the negative and periodic cyclic complexes. If A is a unital algebra then these complexes are quasi-isomorphic to the ones defined above.

2.1 Homology of differential graded algebras

One can easily generalize all the above constructions to the case when A is a differential graded algebra (DGA) with the differential δ (i.e A is a graded algebra and δ is a derivation of degree 1 such that $\delta^2 = 0$).

The action of δ extends to an action on Hochschild chains by the Leibnitz rule:

$$\delta(a_0 \otimes \cdots \otimes a_p) = \sum_{i=1}^{p} (-1)^{\sum_{k<i}(|a_k|+1)+1}(a_0 \otimes \cdots \otimes \delta a_i \otimes \cdots \otimes a_p)$$

The maps b, B, τ are modified to include signs:

$$b(a_0 \otimes \ldots \otimes a_p) = \sum_{k=0}^{p-1}(-1)^{\sum_{i=0}^{k}(|a_i|+1)+1} a_0 \ldots \otimes a_k a_{k+1} \otimes \ldots a_p \tag{10}$$

$$+ (-1)^{|a_p|+(|a_p|+1)\sum_{i=0}^{p-1}(|a_i|+1)} a_p a_0 \otimes \ldots \otimes a_{p-1}$$

$$B(a_0 \otimes \ldots \otimes a_p) = \sum_{k=0}^{p}(-1)^{\sum_{i\leq k}(|a_i|+1)\sum_{i\geq k}(|a_i|+1)} 1 \otimes a_{k+1} \otimes \ldots \otimes a_p \otimes \tag{11}$$

$$\otimes a_0 \otimes \ldots \otimes a_k$$

The complex $C_\bullet(A)$ now becomes the total complex of the double complex with the differential $b + \delta$. The negative and the periodic cyclic complexes are defined as before in terms of the new definition of $C_\bullet(A)$. All the results of this section extend to the differential graded case.

2.2 The Hochschild cochain complex

Let A be a graded algebra with unit over a commutative unital ring k of characteristic zero. A Hochschild d-cochain is a linear map $A^{\otimes d} \to A$. Put, for $d \geq 0$,
$$C^d(A) = C^d(A, A) = \operatorname{Hom}_k(\overline{A}^{\otimes d}, A)$$
where $\overline{A} = A/k \cdot 1$. Put
$$|D| = (\text{degree of the linear map } D) + d$$

Put for cochains D and E from $C^\bullet(A, A)$

$$(D \smile E)(a_1, \ldots, a_{d+e}) = (-1)^{|E| \sum_{i \leq d}(|a_i|+1)} D(a_1, \ldots, a_d) \times$$
$$\times E(a_{d+1}, \ldots, a_{d+e});$$
$$(D \circ E)(a_1, \ldots, a_{d+e-1}) = \sum_{j \geq 0} (-1)^{(|E|+1) \sum_{i=1}^{j}(|a_i|+1)} \times$$
$$\times D(a_1, \ldots, a_j, E(a_{j+1}, \ldots, a_{j+e}), \ldots);$$
$$[D, E] = D \circ E - (-1)^{(|D|+1)(|E|+1)} E \circ D$$

These operations define the graded associative algebra $(C^\bullet(A, A), \smile)$ and the graded Lie algebra $(C^{\bullet+1}(A, A), [\,,\,])$ (cf. [11,32]). Let
$$m(a_1, a_2) = (-1)^{|a_1|} a_1 a_2;$$
this is a 2-cochain of A (not in C^2). Put
$$\delta D = [m, D];$$
$$(\delta D)(a_1, \ldots, a_{d+1}) = (-1)^{|a_1||D|+|D|+1} \times$$
$$\times a_1 D(a_2, \ldots, a_{d+1}) +$$
$$+ \sum_{j=1}^{d} (-1)^{|D|+1+\sum_{i=1}^{j}(|a_i|+1)} D(a_1, \ldots, a_j a_{j+1}, \ldots, a_{d+1})$$
$$+ (-1)^{|D| \sum_{i=1}^{d}(|a_i|+1)} D(a_1, \ldots, a_d) a_{d+1}$$

One has
$$\delta^2 = 0; \quad \delta(D \smile E) = \delta D \smile E + (-1)^{|D|} D \smile \delta E$$
$$\delta[D, E] = [\delta D, E] + (-1)^{|D|+1} [D, \delta E]$$
($\delta^2 = 0$ follows from $[m, m] = 0$).

Thus $C^\bullet(A, A)$ becomes a complex; we will denote it also by $C^\bullet(A)$. The cohomology of this complex is $H^\bullet(A, A)$ or the Hochschild cohomology. We denote it also by $H^\bullet(A)$. The $\smile$ product induces the Yoneda product on $H^\bullet(A, A) = Ext^\bullet_{A \otimes A^0}(A, A)$. The operation $[\,,\,]$ is the Gerstenhaber bracket [32].

If (A, ∂) is a differential graded algebra then one can define the differential ∂ acting on $C^\bullet(A)$ by:

$$\partial D = [\partial, D]$$

Theorem 2.2.1 *[32] The cup product and the Gerstenhaber bracket induce a Gerstenhaber algebra structure on $H^\bullet(A)$ (cf. 4.1 for the definition of a Gerstenhaber algebra).*

For cochains D and D_i define a new Hochschild cochain by the following formula of Gerstenhaber ([32]) and Getzler ([35]):

$$D_0\{D_1, \ldots, D_m\}(a_1, \ldots, a_n) =$$

$$= \sum (-1)^{\sum_{k \leq i_p}(|a_k|+1)(|D_p|+1)} D_0(a_1, \ldots, a_{i_1}, D_1(a_{i_1+1}, \ldots), \ldots,$$

$$D_m(a_{i_m+1}, \ldots), \ldots)$$

Proposition 2.4. *One has*

$$(D\{E_1, \ldots, E_k\})\{F_1, \ldots, F_l\} = \sum (-1)^{\sum_{q \leq i_p}(|E_p|+1)(|F_q|+1)} \times$$

$$\times D\{F_1, \ldots, E_1\{F_{i_1+1}, \ldots,\}, \ldots, E_k\{F_{i_k+1}, \ldots,\}, \ldots,\}$$

The above proposition can be restated as follows. For a cochain D let $D^{(k)}$ be the following k-cochain of $C^\bullet(A)$:

$$D^{(k)}(D_1, \ldots, D_k) = D\{D_1, \ldots, D_k\}$$

Proposition 2.5. *The map*

$$D \mapsto \sum_{k \geq 0} D^{(k)}$$

is a morphism of differential graded algebras

$$C^\bullet(A) \to C^\bullet(C^\bullet(A))$$

2.3 Products on Hochschild and cyclic complexes

Unless otherwise specified, the reference for this subsection is [44].

Product and coproduct; the Künneth exact sequence

For an algebra A define the shuffle product

$$\text{sh} : C_p(A) \otimes C_q(A) \to C_{p+q}(A) \qquad (12)$$

as follows.

$$(a_0 \otimes \ldots \otimes a_p) \otimes (c_0 \otimes \ldots \otimes c_q) = a_0 c_0 \otimes \text{sh}_{pq}(a_1, \ldots, a_p, c_1, \ldots, c_q) \qquad (13)$$

where

$$\text{sh}_{pq}(x_1, \ldots, x_{p+q}) = \sum_{\sigma \in \text{Sh}(p,q)} \text{sgn}(\sigma) x_{\sigma^{-1}1} \otimes \ldots \otimes x_{\sigma^{-1}(p+q)} \qquad (14)$$

and

$$\text{Sh}(p,q) = \{\sigma \in \Sigma_{p+q} | \sigma 1 < \ldots < \sigma p;\ \sigma(p+1) < \ldots < \sigma(p+q)\}$$

In the graded case, $\text{sgn}(\sigma)$ gets replaced by the sign computed by the following rule: in all transpositions, the parity of a_i is equal to $|a_i| + 1$ if $i > 0$, and similarly for c_i. A transposition contributes a product of parities.

The shuffle product is not a morphism of complexes unless A is commutative. It defines, however, an exterior product as shown in the following theorem. For two unital algebras A and C, let i_A, i_C be the embeddings $a \mapsto a \otimes 1$, resp. $c \mapsto 1 \otimes c$ of A, resp. C, to $A \otimes C$. We will use the same notation for the embeddings that i_A, i_C induce on all the chain complexes considered by us.

Theorem 2.3.1 *For two unital algebras A and C the composition*

$$C_p(A) \otimes C_q(C) \xrightarrow{i_A \otimes i_C} C_p(A \otimes C) \otimes C_q(A \otimes C) \xrightarrow{\text{sh}} C_{p+q}(A \otimes C)$$

defines a quasi-isomorphism

$$\text{sh} : C_\bullet(A) \otimes C_\bullet(C) \to C_\bullet(A \otimes C)$$

To extend this theorem to cyclic complexes, define

$$\text{sh}' : C_p(A) \otimes C_q(A) \to C_{p+q+2}(A) \qquad (15)$$

as follows.

$$(a_0 \otimes \ldots \otimes a_p) \otimes (c_0 \otimes \ldots \otimes c_q) \mapsto 1 \otimes \text{sh}'_{p+1,\,q+1}(a_0, \ldots, a_p, c_0, \ldots, c_q) \qquad (16)$$

where

$$\text{sh}'_{p+1,q+1}(x_0, \ldots, x_{p+q+1}) = \sum_{\sigma \in \text{Sh}'(p+1,q+1)} \text{sgn}(\sigma) x_{\sigma^{-1}0} \otimes \ldots \otimes x_{\sigma^{-1}(p+q+1)}$$

$$(17)$$

and Sh$'(p+1, q+1)$ is the set of all permutations $\sigma \in \Sigma_{p+q+2}$ such that $\sigma 0 < \ldots < \sigma p$, $\sigma(p+1) < \ldots < \sigma(p+q+1)$, and $\sigma 0 < \sigma(p+1)$.

Now define (15) to be the composition

$$C_p(A) \otimes C_q(C) \xrightarrow{i_A \otimes i_C} C_p(A \otimes C) \otimes C_q(A \otimes C) \xrightarrow{\text{sh}'} C_{p+q+2}(A \otimes C)$$

In the graded case, the sign rule is as follows: any a_i has parity $|a_i| + 1$, and similarly for c_i.

Theorem 2.3.2 *The map* sh + ush$'$ *defines a* $k[[u]]$*-linear,* (u)*-adically continuous quasi-isomorphism*

$$(C_\bullet(A) \otimes C_\bullet(C))[[u]] \to CC_\bullet^-(A \otimes C)$$

as well as

$$(C_\bullet(A) \otimes C_\bullet(C))[u^{-1}, u]] \to CC_\bullet^{\text{per}}(A \otimes C)$$

$$(C_\bullet(A) \otimes C_\bullet(C))[u^{-1}, u]]/u(C_\bullet(A) \otimes C_\bullet(C))[[u]] \to CC_\bullet(A \otimes C)$$

(the differentials on the left hand sides are equal to $b \otimes 1 + 1 \otimes b + u(B \otimes 1 + 1 \otimes B)$*).*

Note that the left hand side of the last formula maps to the tensor product $CC_\bullet(A) \otimes CC_\bullet(C) : \Delta(u^{-p} c \otimes c') = (u^{-1} \otimes 1 + 1 \otimes u^{-1})^p c \otimes c'$. One checks that this map is an embedding whose cokernel is the kernel of the map $u \otimes 1 - 1 \otimes u$, or $S \otimes 1 - 1 \otimes S$ where S is as in (5) From this we get

Theorem 2.3.3 *There is a long exact sequence*

$$\to HC_n(A \otimes C) \xrightarrow{\Delta} \bigoplus_{p+q=n} HC_p(A) \otimes HC_q(C) \xrightarrow{S \otimes 1 - 1 \otimes S}$$

$$\bigoplus_{p+q=n-2} HC_p(A) \otimes HC_q(C) \xrightarrow{\times} HC_{n-1}(A \otimes C) \xrightarrow{\Delta}$$

2.4 Pairings between chains and cochains

For a graded algebra A, for $D \in C^d(A, A)$, define

$$i_D(a_0 \otimes \ldots \otimes a_n) = (-1)^{|D| \sum_{i \leq d}(|a_i|+1)} a_0 D(a_1, \ldots, a_d) \otimes a_{d+1} \otimes \ldots \otimes a_n \quad (18)$$

Proposition 2.6.

$$[b, i_D] = i_{\delta D}$$

$$i_D i_E = (-1)^{|D||E|} i_{E \smile D}$$

Now, put

$$L_D(a_0 \otimes \ldots \otimes a_n) = \sum_{k=1}^{n-d} \epsilon_k a_0 \otimes \ldots \otimes D(a_{k+1}, \ldots, a_{k+d}) \otimes \ldots \otimes a_n + \quad (19)$$

$$\sum_{k=n+1-d}^{n} \eta_k D(a_{k+1}, \ldots, a_n, a_0, \ldots) \otimes \ldots \otimes a_k$$

(The second sum in the above formula is taken over all cyclic permutations such that a_0 is inside D). The signs are given by

$$\epsilon_k = (|D|+1) \sum_{i=0}^{k} (|a_i|+1)$$

and

$$\eta_k = |D| + 1 + \sum_{i \leq k} (|a_i|+1) \sum_{i \geq k} (|a_i|+1)$$

Proposition 2.7.

$$[L_D, L_E] = L_{[D,E]}$$
$$[b, L_D] + L_{\delta D} = 0$$
$$[L_D, B] = 0$$

Now let us extend the above operations to the cyclic complex. Define

$$S_D(a_0 \otimes \ldots \otimes a_n) = \sum_{j \geq 0;\ k \geq j+d} \epsilon_{jk} 1 \otimes a_{k+1} \otimes \ldots a_0 \otimes \ldots \otimes \quad (20)$$

$$D(a_{j+1}, \ldots, a_{j+d}) \otimes \ldots \otimes a_k$$

(The sum is taken over all cyclic permutations; a_0 appears to the left of D). The signs are as follows:

$$\epsilon_{jk} = |D| \left(|a_0| + \sum_{i=1}^{n} (|a_i|+1) \right) + (|D|+1) \sum_{j+1}^{k} (|a_i|+1) + \sum_{i \leq k} (|a_i|+1) \sum_{i \geq k} (|a_i|+1)$$

As we will see later, all the above operations are partial cases of a unified algebraic structure for chains and cochains; the sign rule for this unified construction will be explained in 2.5.

Proposition 2.8. ([58])

$$[b + uB, i_D + uS_D] - i_{\delta D} - uS_{\delta D} = L_D$$

Proposition 2.9. ([21]) There exists a linear transformation $T(D, E)$ of the Hochschild chain complex, bilinear in $D, E \in C^\bullet(A, A)$, such that

$$[b + uB, T(D, E)] - T(\delta D, E) - (-1)^{|D|} T(D, \delta E) =$$
$$= [L_D, i_E + uS_E] - (-1)^{|D|+1}(i_{[D,E]} + uS_{[D,E]})$$

2.5 A_∞ structure on $C_\bullet(C^\bullet(A))$

All the pairings described above are in fact different parts of the same algebraic structure that we are going to describe next.

Recall that an A_∞ algebra is a graded vector space $\mathcal{C}$ together with a Hochschild cochain m of total degree 1,

$$m = \sum_{n=1}^{\infty} m_n$$

where $m_n : \mathcal{C}^{\otimes n} \to \mathcal{C}$ is a Hochschild n-cochain and

$$[m, m] = 0$$

To define A_∞ modules over A_∞ algebras, note that for a graded space $\mathcal{M}$, the Gerstenhaber bracket $[\,,\,]$ can be extended to the space

$$\text{Hom}(\mathcal{C}^{\otimes \bullet}, \mathcal{C}) \oplus \text{Hom}(\mathcal{M} \otimes \mathcal{C}^{\otimes \bullet}, \mathcal{M})$$

For a graded space $\mathcal{M}$, a structure of an A_∞ module over an A_∞ algebra $\mathcal{C}$ on $\mathcal{M}$ is a cochain

$$\mu = \sum_{n=1}^{\infty} \mu_n$$

$$\mu_n \in Hom(\mathcal{M} \otimes \mathcal{C}^{\otimes n-1}, \mathcal{M})$$

such that

$$[m + \mu, m + \mu] = 0$$

Consider the Hochschild cochain complex of a graded algebra A as a differential graded associative algebra $(C^\bullet(A), \smile, \delta)$. Consider the Hochschild *chain* complex of this differential graded algebra. The total differential in this complex is $b + \delta$; the degree of a chain is given by

$$|D_0 \otimes \ldots \otimes D_n| = |D_0| + \sum_{i=1}^{n}(|D_i| + 1)$$

where D_i are Hochschild cochains.

The complex $C_\bullet(C^\bullet(A))$ contains the Hochschild cochain complex $C^\bullet(A)$ as a subcomplex (of zero-chains) and has the Hochschild chain complex $C_\bullet(A)$ as a quotient complex:

$$C^\bullet(A) \xrightarrow{i} C_\bullet(C^\bullet(A)) \xrightarrow{\pi} C_\bullet(A)$$

(this sequence is not exact in any sense). The projection on the right splits if A is commutative. If not, $C_\bullet(A)$ is naturally a graded subspace but not a subcomplex.

Theorem 2.5.1 *There is an A_∞ structure on $C_\bullet(C^\bullet(A))[[u]]$ such that:*

- *All m_n are $k[[u]]$-linear, (u)-adically continuous*
- *$m_1 = b + \delta + uB$*
  For $x, y \in C_\bullet(A)$:
- *$(-1)^{|x|} m_2(x, y) = (\operatorname{sh} + u \operatorname{sh}')(x, y)$*
  For $D, E \in C^\bullet(A)$:
- *$(-1)^{|D|} m_2(D, E) = D \smile E$*
- *$m_2(1 \otimes D, 1 \otimes E) + (-1)^{|D||E|} m_2(1 \otimes E, 1 \otimes D) = (-1)^{|D|} 1 \otimes [D, E]$*
- *$m_2(D, 1 \otimes E) + (-1)^{(|D|+1)|E|} m_2(1 \otimes E, D) = (-1)^{|D|+1} [D, E]$*

Theorem 2.5.2 *On $C_\bullet(A)[[u]]$, there exists a structure of an A_∞ module over the A_∞ algebra $C_\bullet(C^\bullet(A))[[u]]$ such that:*

- *All μ_n are $k[[u]]$-linear, (u)-adically continuous*
- *$\mu_1 = b + uB$ on $C_\bullet(A)[[u]]$*
  For $a \in C_\bullet(A)[[u]]$:
- *$\mu_2(a, D) = (-1)^{|a||D|+|a|}(i_D + u S_D)a$*
- *$\mu_2(a, 1 \otimes D) = (-1)^{|a||D|} L_D a$*
  For $a, x \in C_\bullet(A)[[u]]$: $(-1)^{|a|} \mu_2(a, x) = (\operatorname{sh} + u \operatorname{sh}')(a, x)$

Here is an explicit description of the above A_∞ structures. We define for $n \geq 2$

$$m_n = m_n^{(1)} + m_n^{(2)} + u m_n^{(3)}$$

where, for

$$a^{(k)} = D_0^{(k)} \otimes \ldots \otimes D_{N_k}^{(k)},$$

$$m_n^{(1)} = 0$$

for $n \geq 3$;

$$m_2^{(1)}(a^{(1)}, a^{(2)}) = (-1)^{|a^{(1)}|} \sum \pm D_0^{(1)} \smile D_0^{(2)}\{\underline{}\} \otimes \ldots \otimes$$

$$\otimes D_1^{(2)}\{\underline{}\} \otimes \ldots \otimes D_{N_2}^{(2)}\{\underline{}\} \otimes \ldots$$

The space designated by $\underline{}$ is filled with $D_1^{(1)}, \ldots, D_{N_1}^{(1)}$ whose order is preserved. The sign rule is as follows: the parity of $D_j^{(i)}$ is $|D_j^{(i)}|$ for $j = 0$ and $|D_j^{(i)}| + 1$ otherwise.

$$m_n^{(2)}(a^{(1)}, \ldots, a^{(n)}) =$$

$$(-1)^{\sum_{i=1}^{n-1} |a_i| + n} \sum \pm D_{N_n}^{(n)} \{\underline{\ldots}, D_0^{(1)}, \underline{\ldots}, D_0^{(n-1)}, \underline{\ldots}\} \smile D_0^{(n)}$$

$$\otimes \underline{\ldots} \otimes D_1^{(n)}\{\underline{}\} \otimes \underline{\ldots} \otimes D_{N_n-1}^{(n)}\{\underline{}\} \otimes \underline{\ldots}$$

The space designated by $\underline{}$ is filled with $D_i^{(j)}$ for $j < n$ in such a way that:

- the cyclic order of each group $D_0^{(k)}, \ldots, D_{N_k}^{(k)}$ is preserved
- $D_0^{(1)}, \ldots, D_0^{(n-1)}$ are all inside the braces in $D_{N_n}^{(n)}\{\ \}$
- $D_0^{(i)}$ is to the left of $D_0^{(j)}$ for $i < j$
- any cochain $D_j^{(i)}$ may contain some of its neighbors on the right inside the braces, provided that all of these neighbors are of the form $D_q^{(p)}$ with $p < i$

The parity of $D_j^{(i)}$ is $|D_j^{(i)}|$ if $i = n$ and $j = 0$; it is $|D_j^{(i)}| + 1$ otherwise.

$$m_n^{(3)}(a^{(1)}, \ldots, a^{(n)}) = (-1)^{n+1} \sum \pm 1 \otimes \ldots \otimes D_0^{(0)} \otimes \ldots \otimes D_0^{(n)} \otimes \ldots$$

The space designated by _ is filled with $D_i^{(j)}$ for in such a way that:

- the cyclic order of each group $D_0^{(k)}, \ldots, D_{N_k}^{(k)}$ is preserved
- $D_0^{(i)}$ is to the left of $D_0^{(j)}$ for $i < j$
- any cochain $D_j^{(i)}$ may contain some of its neighbors on the right inside the braces, provided that all of these neighbors are of the form $D_q^{(p)}$ with $p < i$. The parity of $D_j^{(i)}$ is always $|D_j^{(i)}| + 1$.

The above formulas provide an A_∞ structure from Theorem 2.5.1. To obtain a structure of an A_∞ module from Theorem 2.5.2, one has to assume that all $D_j^{(1)}$ are elements of A and to replace braces $\{\ \}$ by the usual parentheses $(\ )$ symbolizing evaluation of a multi-linear map at elements of A.

Remark 2.10. Let A be a commutative algebra. Then $C_\bullet(A)[[u]]$ is not only a subcomplex but an A_∞ subalgebra of $C_\bullet(C^\bullet(A))[[u]]$. This A_∞ structure on $C_\bullet(A)[[u]]$ was introduced in [36].

The corresponding binary product was defined by Hood and Jones [39]. One can define it for any algebra A, commutative or not. If A is not commutative then this product is not compatible with the differential. Nevertheless, we will use it in 3.5.

3 The cyclic complex $C_\bullet^\lambda$

3.1 Definition

Recall the original definition of the cyclic complex from [13], [72]. Let $\tau = \tau_p$ denote the endomorphism of $A^{\otimes k(p+1)}$ given by the formula

$$\tau(a_0 \otimes \cdots \otimes a_p) = (-1)^p a_p \otimes a_0 \cdots \otimes a_{p-1}$$

Let

$$C_p^\lambda(A) = A^{\otimes k(p+1)}/\mathrm{Im}(\mathrm{id} - \tau) \ . \tag{1}$$

Define
$$b' : C_p(A) \to C_{p-1}(A)$$
$$a_0 \otimes \cdots \otimes a_p \mapsto \sum_{i=0}^{p-1} (-1)^i a_0 \otimes \cdots \otimes a_i a_{i+1} \otimes \cdots \otimes a_p$$

and
$$N : C_p(A) \to C_{p-1}(A)$$
$$N = \mathrm{id} + \tau + \ldots \tau^{p-1}$$

One has
$$b(\mathrm{id} - \tau) = (\mathrm{id} - \tau)b'; \quad b'N = Nb. \tag{2}$$

Therefore the differential b descends to a map
$$b : C_p^\lambda(A) \to C_{p-1}^\lambda(A) \ .$$

The complex $C_\bullet^\lambda(A)$ is quasi-isomorphic to $CC_\bullet(A)$. (Sketch of the proof: in the cyclic double complex $CC_\bullet(A)$ one can replace $C_\bullet(A)$ by the cone of $\mathrm{id} - \tau$ and B by N. One gets a double complex which is quasi-isomorphic to $CC_\bullet$ and which is in turn quasi-isomorphic to $C_\bullet^\lambda$. Cf. [44] for more detail).

Lemma 3.1. *For algebras without unit, the cyclic homology defined in Remark 2.3 is isomorphic to the homology of the complex $C_\bullet^\lambda(A)$.*

3.2 The reduced cyclic complex

Let $\overline{A} = A/k$ and let
$$\overline{C}_p^\lambda(A) = \overline{A}^{\otimes p+1}/\mathrm{Im}(\mathrm{id} - \tau) \ . \tag{3}$$

It is easy to see that the diferential b descends to $\overline{C}_\bullet^\lambda(A)$. We denote the homology of the complex $\overline{C}_\bullet^\lambda(A)$ by $\overline{HC}_\bullet(A)$.

Proposition 3.2.1 *There is an exact triangle*
$$C_\bullet^\lambda(k) \to C_\bullet^\lambda(A) \to \overline{C}_\bullet^\lambda(A) \to C_\bullet^\lambda(k)[1] \ . \tag{4}$$

Cf. [44] for the proof. (Note that the above proposition could be deduced from the Hochschild-Serre spectral sequence
$$E_{pq}^2 = H_p(\mathfrak{gl}(A), \mathfrak{gl}(k)) \otimes H_q(\mathfrak{gl}(k))$$
converging to $H_{p+q}(\mathfrak{gl}(A))$ and from theorem 3.3.1).

3.3 Relation to Lie algebra homology

For any DG Lie algebra $\mathfrak{g}$, let $C_\bullet(\mathfrak{g})$ be the standard Chevalley-Eilenberg complex of chains of $\mathfrak{g}$ with coefficients in the trivial module k. For a DG Lie subalgebra $\mathfrak{h}$ acting reductively on $\mathfrak{g}$, let $C_\bullet(\mathfrak{g})_\mathfrak{h}$ be the complex of coinvariants $C_\bullet(\mathfrak{g})$ with respect to the adjoint action of $\mathfrak{h}$, and let $C_\bullet(\mathfrak{g}, \mathfrak{h})$ be the complex of relative chains.

For any DG algebra (A, δ) over k let $\mathfrak{gl}(A) = \varinjlim_n \mathfrak{gl}_n(A)$ denote the Lie algebra of finite matrices. Note that $\mathfrak{gl}(k)$ is a DG Lie subalgebra of $\mathfrak{gl}(A)$. Theorem 3.3.1 below identifies the cyclic complex of A, resp. the relative cyclic complex of A, with the subcomplex of primitive elements of the DG coalgebra of coinvariants $C_\bullet(\mathfrak{gl}(A))_{\mathfrak{gl}(k)}$, resp. $C_\bullet(\mathfrak{gl}(A), \mathfrak{gl}(k))$.

Let E_{pq}^a denote the elementary matrix with $\left(E_{pq}^a\right)_{pq} = a$ and other entries equal to zero.

Theorem 3.3.1 *The map*

$$A^{\otimes p+1} \to \bigwedge^{p+1} \mathfrak{gl}(A)$$

$$a_0 \otimes \cdots \otimes a_p \mapsto E_{01}^{a_0} \wedge E_{12}^{a_1} \wedge \cdots \wedge E_{p-1,0}^{a_p}$$

induces isomorphisms of complexes

$$C_\bullet^\lambda(A) \to \operatorname{Prim} C_\bullet(\mathfrak{gl}(A))_{\mathfrak{gl}(k)}[1]$$

$$\overline{C}_\bullet^\lambda(A) \to \operatorname{Prim} C_\bullet(\mathfrak{gl}(A), \mathfrak{gl}(k))[1]$$

The proof is contained in [44, 45, 71].

3.4 The connecting morphism

Here we give an explicit formula for the connecting morphism ∂ of the exact triangle from Proposition 3.2.1. For a DGA (A, δ) we denote by $T(A)$ the tensor algebra of A and by j a k-linear map $A \to k$ such that $j(1) = 1$.

Let $\rho : T(A) \to k$ denote the map defined by

$$\rho(a) = j(\delta a) \text{ for } a \in A,$$
$$\rho(a_1 \otimes a_2) = j(a_1)j(a_2) - j(a_1 a_2) \text{ for } a_1, a_2 \in A, \text{ and}$$
$$\rho = 0 \text{ on } A^{\otimes p} \text{ for } p \neq 1, 2.$$

For $a_i \in A$, $a_0 \otimes \cdots \otimes a_p \in A^{\otimes p+1}$ with $\sum_i(|a_i|+1) - 1 = 2n+1$ let

$$\operatorname{br}_{2n+1}(a_0 \otimes \cdots \otimes a_p) =$$
$$\frac{1}{n!} \sum_{i=0}^{p} (-1)^{\sum_{k<i}(|a_k|+1) \sum_{k \geq i}(|a_k|+1)} (\rho^{\otimes n})(a_i \otimes \cdots a_0 \otimes \cdots \otimes a_{i-1}). \quad (5)$$

This defines a map $\mathrm{br}_{2n+1}: \overline{C}^\lambda_{2n+1}(A) \to k$ (cf. [6]). Let $\mathrm{br}_{2n} = 0$ on $\overline{C}^\lambda_{2n}(A)$.
Let $1^{(n+1)} = n!(n+1)! \cdot 1^{\otimes 2n+1} \in C^\lambda_{2n}(k)$. Let $\mathrm{Br}^A_{2n+1} = \mathrm{br}_{2n+1} \cdot 1^{(n)}$, $\mathrm{Br}^A_{2n} = 0$, $\mathrm{Br}^A = \sum_p \mathrm{Br}^A_p$.

One gets the map of graded spaces

$$\mathrm{Br}^A : \overline{C}^\lambda_\bullet(A) \to C^\lambda_\bullet(k)[1]$$

which is, in fact, a map of complexes as follows from the alternative description given below.

The splitting j induces the splitting $\rho : \mathfrak{gl}(A) \to \mathfrak{gl}(k)$ of the inclusion $\mathfrak{gl}(k) \hookrightarrow \mathfrak{gl}(A)$. Let P_n denote the invariant polynomial $X \mapsto \frac{1}{n!}tr(X^n)$ on $\mathfrak{gl}(k)$. Put

$$R(\rho) = (\partial^{\mathrm{Lie}} + \delta)R(\rho) + \frac{1}{2}[R(\rho), R(\rho)] \qquad (6)$$

Then br_{2n+1} is the Chern character cochain $P_n(R(\rho))$ composed with the morphism which is inverse to the last isomorphism from Theorem 3.3.1.

Lemma 3.2. *The restriction of the pairing with $P_n(R(\rho))$ to the subspace of all primitive elements (see Theorem 3.3.1) coincides with the map br_{2n+1}.*

The proof is straightforward.

Corollary 3.4.1 *The map Br^A is a morphism of complexes.*

Proposition 3.4.2 *The morphism $\mathrm{Br}^A : \overline{C}^\lambda_\bullet(A) \to C^\lambda_\bullet(k)[1]$ represents the connecting morphism in the triangle (4).*

Explicit formula for the product $HC_p \otimes HC_q \to HC_{p+q+1}$

Here we give an explicit formula for the product $\times$ from Theorem 2.3.3 in one of the realizations of the cyclic complex. Note first that, because of (2), the map N induces an isomorphism

$$C^\lambda(A) \simeq (\mathrm{Ker}(\mathrm{id} - \tau), b') \qquad (7)$$

where in the right hand side $\mathrm{id} - \tau$ is considered as an operator on $A^{\otimes(\bullet+1)}$.

Proposition 3.3. *If one identifies $C^\lambda_\bullet$ with the right hand side of (7), then the shuffle product*

$$(a_0 \otimes \ldots \otimes a_p) \times (c_0 \otimes \ldots \otimes c_q) = \mathrm{sh}_{p+1,q+1}((a_0, \ldots \otimes a_p, c_0, \ldots, c_q)$$

is a morphism of complexes

$$\times : C^\lambda_\bullet(A) \otimes C^\lambda_\bullet(C) \to C^\lambda_{\bullet+1}(A \otimes C) \qquad (8)$$

which induces on homology the $\times$ product from Theorem 2.3.3.

By the same construction one defines the product

$$\times : \overline{C}_\bullet^\lambda(A) \otimes \overline{C}_\bullet^\lambda(C) \to \overline{C}_\bullet^\lambda(A \otimes C)$$

on the reduced cyclic homology. Cf. [44] for the explicit formula for the $\times$ products at the level of complexes $CC_\bullet$.

3.5 Other operations on the cyclic complexes

Here we recall, in a modified version, some results from [53].

Denote by $\mathfrak{g}_A^\bullet$ the DG Lie algebra $C^{\bullet+1}(A,A)$ with the Gerstenhaber bracket and the Hochschild differential δ. Let $\mathfrak{g}_A^\bullet[\epsilon] = \mathfrak{g}_A^\bullet + \mathfrak{g}_A^\bullet \epsilon$, $\epsilon^2 = 0$, $\deg \epsilon = 1$. Let u be a formal parameter of degree 2 and consider the graded algebra $U(\mathfrak{g}_A^\bullet[\epsilon])[u]$ with the differential $u \cdot \frac{\partial}{\partial \epsilon} + \delta$ (recall that δ is the differential in $\mathfrak{g}_A^\bullet$).

The motivation for the theorem below is the following. If $A = C^\infty(M)$ then $HC_\bullet^-(A)$ is isomorphic to the cohomology of the complex $(\Omega^\bullet(M)[[u]], ud)$; the cohomology of $\mathfrak{g}_A^\bullet$ is the graded Lie algebra $\mathfrak{g}_M^\bullet$ of multivector fields on M; one can define an action of $\mathfrak{g}_M^\bullet[\epsilon][u]$ on $HC_\bullet^-(A)$: for two multivector fields X, Y the action of $X + \epsilon Y$ is given by $L_X + \iota_Y$ where ι_Y is the contraction operator and $L_X = [d, \iota_X]$.

We would like to have a noncommutative analog of the above action. In fact, because of Theorem 4.4.1, we know that there is an L_∞ action of $\mathfrak{g}_A^\bullet[\epsilon][u]$ on $CC_\bullet^-(A)$, i.e., a DGLA which is quasi-isomorphic to the former acts on a complex quasi-isomorphic to the latter. However, the proof of Theorem 4.4.1 is inexplicit, and we need explicit operations for index-theoretical applications. Below we provide an explicit formula for an action of the *complex* $U(\mathfrak{g}_A^\bullet[\epsilon])[u]$ on $CC_\bullet^-(A)$.

Denote by $\mathcal{E}_A^\bullet$ the DGA $C^\bullet(A)$ with the cup product and the differential δ. Let $\star$ be the Hood - Jones product on $CC_\bullet^-(\mathcal{E}_A^\bullet)$ (Remark 2.10). In other words,

$$a \star b = (-1)^{|a|}(m_2^{(1)}(a,b) + u m_2^{(3)}(a,b))$$

where $m_i^{(j)}$ are defined in 2.5 and applied here to the subspace $CC_\bullet^-(C^0(\mathcal{E}))$ of $CC_\bullet^-(C^\bullet(\mathcal{E}))$

Theorem 3.5.1 *Formula*

$$(\epsilon D_1 \cdot \cdots \cdot \epsilon D_m) \bullet \alpha = I(D_1, \ldots, D_m)\alpha =$$
$$(-1)^{|\alpha|\sum_{i=1}^m (|D_i|+1)} \frac{1}{m!} \sum_\sigma \epsilon_\sigma \alpha \bullet (D_{\sigma_1} \star (D_{\sigma_2} \star (\ldots \star D_{\sigma_m}))\ldots)$$

defines morphisms of complexes of $k[u]$-modules

$$U(\mathfrak{g}_A^\bullet[\epsilon])[[u]] \otimes_{U(\mathfrak{g}_A^\bullet)[u]} CC_\bullet^-(A) \to CC_\bullet^-(A)$$
$$U(\mathfrak{g}_A^\bullet[\epsilon])[u^{-1}, u]] \otimes_{U(\mathfrak{g}_A^\bullet)[u, u^{-1}]} CC_\bullet^{per}(A) \to CC_\bullet^{per}(A)$$

(the signs ϵ_σ are computed according to the rule under which the parity of any D_i is $|D_i| + 1$). Let u be a formal parameter of degree two. Consider the differential graded algebra $A[\eta] = A + A\eta$, $\deg\eta = -1$, $\eta^2 = 0$ with the differential $\frac{\partial}{\partial \eta}$. Consider the complex $\overline{C}^\lambda_\bullet(A[\eta])[u]$ with the differential $\frac{\partial}{\partial \eta} + u \cdot b$.

Theorem 3.5.2 *There exist natural pairings of $k[[u]]$-modules*

$$\bullet : \overline{C}^\lambda_{\bullet-1}(A[\eta])[[u]] \otimes CC^-_{-\bullet}(A) \to CC^-_{-\bullet}(A)$$

$$\bullet : \overline{C}^\lambda_{\bullet-1}(A[\eta])[[u, u^{-1}] \otimes CC^{per}_{-\bullet}(A) \to CC^{per}_{-\bullet}(A)$$

such that:

1. $\eta^{\otimes m} \bullet ? = \frac{1}{(m-1)!} Id$ *for $m > 0$.*
2. *For $x_i \in A$ the operation $(x_1 \otimes \cdots \otimes x_p) \bullet ?$ sends $C_N(A)$ to $\sum_{i,j \geq 0} C_{N-p+i}(A)u^j$.*
3. *The component of $(x_1 \otimes \cdots \otimes x_p) \bullet (a_0 \otimes \cdots \otimes a_N)$ in $C_{N-p}[[u]]$ is equal to*

$$\frac{1}{p!} \sum_{i=1}^{p}(-1)^{i(p-1)} a_0[x_{i+1}, a_1][x_{i+2}, a_2] \cdots [x_i, a_p] \otimes a_{p+1} \otimes \cdots \otimes a_N$$

4. $(x_1 \otimes \cdots \otimes x_p) \bullet 1 = \sum_{i=1}^{p}(-1)^{i(p-1)} 1 \otimes x_{i+1} \otimes x_{i+2} \otimes \cdots \otimes x_i.$

Note that the formula (4) above defines the map from $\overline{C}^\lambda_\bullet(A)$ to $Ker(B : C_\bullet(A) \to C_{\bullet+1}(A))$ (one can show that this map is a quasi-isomorphism). Clearly, the kernel above embeds into $CC^-_\bullet(A)$. The above theorem shows that this embedding extends to a pairing $\bullet$.

The complex $\overline{C}^\lambda_{\bullet-1}(A[\eta])[[u]]$ is very simple at the level of homology; it is quasi-isomorphic to $k[[u]]$. Therefore the pairing $\bullet$ does not define any new homological operations. It is, however, very important at the level of chains, as one sees in [4].

3.6 Rigidity of periodic cyclic homology

Let A be an associative algebra over a ring k of characteristic zero. Let I be a nilpotent two-sided ideal of A. Denote $A_0 = A/I$.

Theorem 3.6.1 *(Goodwillie) The natural map $CC^{per}_\bullet(A) \to CC^{per}_\bullet(A/I)$ is a quasi-isomorphism.*

Proof. Fix a k-linear isomorphism $A \simeq A_0 \oplus I$. Denote by $*$ the image of the following multiplication under this isomorphism: the product is the product on A if one of the rguments is in A_0, and $I^2 = 0$. Put $\lambda(a_1, a_2) = a_1 a_2 - a_1 * a_2$; then λ is an element of $\mathfrak{g}^1_A = C^2(A)$ such that

$$\delta\lambda + \frac{1}{2}[\lambda, \lambda] = 0.$$

Since λ takes values in I, the operation

$$\chi = \sum_{m=0}^{\infty} \frac{1}{m!} I(\lambda, \ldots, \lambda) \qquad (9)$$

is well-defined on $CC_\bullet^{\mathrm{per}}(A)$ (where $I(\lambda, \ldots, \lambda)$ are the operators from theorem 3.5.1). One checks that χ is indeed a morphism of complexes ([53]). It is easy to check that this morphism is invertible.

4 Noncommutative differential calculus

4.1 Gerstenhaber algebras

Let k be the ground ring of characteristic zero. A *Gerstenhaber algebra* is a graded space $\mathcal{V}^\bullet$ together with

- A graded commutative associative algebra structure on $\mathcal{V}^\bullet$;
- a graded Lie algebra structure on $\mathcal{V}^{\bullet+1}$ such that

$$[a, bc] = [a, b]c + (-1)^{(|a|-1)|b|} b[a, c]$$

Example 4.1. Let M be a smooth manifold. Then

$$\mathcal{V}_M^\bullet = \wedge^\bullet T_M$$

is a sheaf of Gerstenhaber algebras.

The product is the exterior product, and the bracket is the Schouten bracket. We denote by $\mathcal{V}^\bullet(M)$ the Gerstenhaber algebra of global sections of this sheaf.

Example 4.2. Let $\mathfrak{g}$ be a Lie algebra. Then

$$C_\bullet(\mathfrak{g}) = \wedge^\bullet \mathfrak{g}$$

is a Gerstenhaber algebra.

The product is the exterior product, and the bracket is the unique bracket which turns $C_\bullet(\mathfrak{g})$ into a Gerstenhaber algebra and which is the Lie bracket on $\mathfrak{g} = \wedge^1(\mathfrak{g})$.

4.2 The Gerstenhaber algebra $\mathcal{V}^\bullet(A)$

Below is the theorem from [64].

Theorem 4.2.1 *For every associative algebra A there exists a dg Gerstenhaber algebra $\mathcal{V}^\bullet(A)$ such that:*

- There is a quasi-isomorphism of dg Lie algebras
$$\mathcal{V}^\bullet(A)[1] \to C^\bullet(A)[1]$$
- The above quasi-isomorphism induces an isomorphism of Gerstenhaber algebras
$$H^\bullet(\mathcal{V}^\bullet(A)) \to H^\bullet(A)$$
where the Gerstenhaber structure on the right hand side is the standard one from 2.2.
- For $A = C^\infty(M)$ there is a quasi-isomorphism of dg Gerstenhaber algebras
$$\mathcal{V}^\bullet(A) \to \mathcal{V}^\bullet(M)$$

4.3 Calculi

Definition 4.3. *A precalculus is a pair of a Gerstenhaber algebra $\mathcal{V}^\bullet$ and a graded space $\Omega^\bullet$ together with*

- *a structure of a graded module over the graded commutative algebra $\mathcal{V}^\bullet$ on $\Omega^{-\bullet}$ (corresponding action is denoted by i_a, $a \in \mathcal{V}^\bullet$);*
- *a structure of a graded module over the graded Lie algebra $\mathcal{V}^{\bullet+1}$ on $\Omega^{-\bullet}$ (corresponding action is denoted by L_a, $a \in \mathcal{V}^\bullet$) such that*

$$[L_a, i_b] = i_{[a,b]}$$

and

$$L_{ab} = L_a\, i_b + (-1)^{|a|} i_a\, L_b$$

Definition 4.4. *A calculus is a precalculus together with an operator d of degree 1 on $\Omega^\bullet$ such that $d^2 = 0$ and*

$$[d, i_a] = L_a.$$

Example 4.5. For any manifold one defines a calculus $\mathrm{Calc}(M)$ with $\mathcal{V}^\bullet$ being the algebra of multivector fields, $\Omega^\bullet$ the space of differential forms, and d the De Rham differential. The operator i_a is the contraction of a form by a multivector field.

Example 4.6. For any associative algebra A one defines a calculus $\mathrm{Calc}_0(A)$ by putting $\mathcal{V}^\bullet = H^\bullet(A, A)$ and $\Omega^\bullet = H_\bullet(A, A)$. The five operations from Definition 4.4 are the cup product, the Gerstenhaber bracket, the pairings i_D and L_D, and the differential B, as in 2.4. The fact that it is indeed a calculus follows from Theorem 2.9.

A differential graded (dg) calculus is a calculus with extra differentials δ of degree 1 on $\mathcal{V}^\bullet$ and b of degree -1 on $\Omega^\bullet$ which are derivations with respect to all the structures.

4.4 The calculus Calc(A)

Theorem 4.4.1 *[66] For any associative algebra A one can define a dg calculus $\mathrm{Calc}(A)$ such that:*

1). As dg Lie algebras, $\mathcal{V}^{\bullet+1}(A)$ is quasi-isomorphic to the Hochschild cochain complex $C^{\bullet+1}(A)$. As dg modules over the dg Lie algebra $\mathcal{V}^{\bullet+1}(A)$, $\Omega^{\bullet}(A)[[u]]$ with the differential $\delta + ud$ is quasi-isomorphic to the negative cyclic complex $C_{\bullet}(A)[[u]]$ with the differential $b + uB$.

2). If $A = C^{\infty}(M)$ then there is a quasi-isomorphism of calculi

$$\mathrm{Calc}(A) \to \mathrm{Calc}(M)$$

3). For any A, the calculus $H^{\bullet}(\mathrm{Calc}(A))$ is isomorphic to the calculus $\mathrm{Calc}_0(A)$ from Example 4.6.

Combining 2) and 3), one gets for $A = C^{\infty}(M)$ quasi-isomorphisms

$$C_{\bullet}(A)[[u]] \leftarrow \Omega^{\bullet}(A)[[u]] \to \Omega^{\bullet}(M)[[u]]. \tag{1}$$

Let **1** be the canonical homology class of the left hand side. Its image under the map induced on the cohomology by (1) is equal to

$$\sum u^p \tau_{2p}$$

where τ_{2p} are some De Rham cohomology classes of M. We finish this subsection by stating what is known at this moment about the class

$$\tau = \sum \tau_{2p}.$$

Consider the ring

$$\mathcal{R} = \mathbb{C}[[c_1, c_2, \ldots]]$$

where $|c_i| = 2i$. Let Let

$$\pi : \mathcal{R} \to \mathcal{R}^{\mathrm{even}}$$

be the homomorphism which sends c_i to c_i if i is even and to zero if i is odd.

Theorem 4.4.2 *The class τ is a characteristic class of the tangent bundle of M which is given by a power series $\tau \in \mathcal{R}$ such that*

$$\pi(\tau) = \sqrt{\widehat{A}}$$

Here

$$\widehat{A}(c_1, c_2, \ldots) = \prod_i \frac{x_i}{e^{x_i/2} - e^{-x_i/2}}$$

where c_i is the ith elementary symmetric polynomial of $x_1, x_2, \ldots$.

4.5 Enveloping algebra of a Gerstenhaber algebra

The following construction is motivated by Example 4.5. For a Gerstenhaber algebra $\mathcal{V}^\bullet$, let $Y(\mathcal{V}^\bullet)$ be the associative algebra generated by two sets of generators i_a, L_a, $a \in \mathcal{V}^\bullet$, both i and L linear in a,

$$|i_a| = |a|; \; |L_a| = |a| - 1$$

subject to relations

$$i_a i_b = i_{ab}; \; [L_a, L_b] = L_{[a,b]};$$

$$[L_a, i_b] = i_{[a,b]}; \; L_{ab} = L_a i_b + (-1)^{|a|} i_a L_b$$

The algebra $Y(\mathcal{V}^\bullet)$ is equipped with the differential d of degree one which is defined as a derivation sending i_a to L_a and L_a to zero.

For a smooth manifold M one has a homomorphism

$$Y(\mathcal{V}^\bullet(M)) \to \mathrm{D}(\Omega^\bullet(M))$$

The right hand side is the algebra of differential operators on differential forms on M, and the above homomorphism sends the generators i_a, L_a to corresponding differential operators on forms (cf. Example 4.5). It is easy to see that the above map is in fact an isomorphism.

4.6 Relation to the A_∞ structure on chains

Theorem 4.6.1 ([69]). *There is a A_∞ quasi-isomorphism*

$$Y(\mathcal{V}^\bullet(A)) \to C_\bullet(C^\bullet(A))$$

which extends to a $k[[u]]$-linear, (u)-adically continuous A_∞ quasi-isomorphism

$$(Y(\mathcal{V}^\bullet(A))[[u]], \delta + ud) \to C_\bullet(C^\bullet(A))[[u]]$$

The meaning of the above theorems 4.4.1 and 4.6.1 is as follows. First of all, many of the algebraic structures from the classical calculus appear in the context of noncommutative geometry. This is somewhat counterintuitive because these algebraic structures include two graded commutative multiplications, one on forms and another on multivector fields. The multiplication on forms has no noncommutative analog. The multiplication of multivectors does have one, and this is the least trivial part of noncommutative calculus. The graded commutative algebra structure on $\mathcal{V}^\bullet(A)$, as well as the action of this algebra on $\Omega^\bullet(A)$, exists due to Theorem 4.4.1, but it is highly inexplicit and non-canonical; it depends on a choice of a Drinfeld associator. If one considers less restrictive algebraic structures from classical calculus, more or less anything involving Lie algebras, associative algebras, and modules over them, then one can generalize them to the context of noncommutative geometry in an explicit and canonical way. This generalization is the subject of Theorems 4.6.1 and 3.5.1.

5 Cyclic objects

5.1 Simplicial objects

Denote by Δ the category whose objects are sets $[n] = \{0, 1, \ldots, n\}$, $n \geq 0$, and morphisms are nondecreasing maps. For a category C define a simplicial object of C an a functor from Δ^0 to C where Δ^0 is the category opposite to Δ. Any morphism in Δ can be written as a composition of the following:

1. Embeddings $d_i : [n] \to [n+1]$, $0 \leq i \leq n+1$ (i is not in the image of d_i);
2. Surjections $s_i : [n+1] \to [n]$, $0 \leq i \leq n$ (i is the image of two points under s_i).

A simplicial object of C can be equivalently described as a collection of objects X_n of C, $n \geq 0$, together with morphisms

$$d_i : X_{n+1} \to X_n, \ 0 \leq i \leq n+1 \tag{1}$$

$$s_i : X_n \to X_{n+1}, \ 0 \leq i \leq n \tag{2}$$

subject to

$$d_i d_j = d_{j-1} d_i, \ i < j \tag{3}$$

$$s_i s_j = s_{j+1} s_i, \ i \leq j \tag{4}$$

$$d_i s_j = s_{j-1} d_j, \ i < j; \ d_i s_i = d_{i+1} s_i = \mathrm{id}; \ d_i s_j = s_j d_{i-1}, \ i > j+1 \tag{5}$$

Example 5.1. For a topological space X define $\mathrm{Sing}_n(X)$ to be the set of singular simplices of X. Then $\mathrm{Sing}(X)$ has a natural structure of a simplicial set.

Example 5.2. Let A be a graded algebra. Put $A_n^{\#} = A^{\otimes(n+1)}$. Define

$$d_i(a_0 \otimes \ldots \otimes a_{n+1}) = (-1)^{\sum_{p \leq i} |a_p|} a_0 \otimes \ldots a_i a_{i+1} \otimes \ldots \otimes a_{n+1} \tag{6}$$

for $i \leq n$

$$d_{n+1}(a_0 \otimes \ldots \otimes a_{n+1}) = (-1)^{|a_{n+1}| \sum_{p \leq n} |a_p|} a_{n+1} a_0 \otimes a_1 \otimes \ldots \otimes a_n \tag{7}$$

$$s_i(a_0 \otimes \ldots \otimes a_n) = (-1)^{\sum_{p \leq i} |a_p|} a_0 \otimes \ldots a_i \otimes 1 \otimes \ldots \otimes a_n \tag{8}$$

for $i \leq n$

Proposition 5.3. *The above formulas make $A^{\#}$ a simplicial graded space.*

5.2 Cyclic objects

A cyclic object of a category C is a simplicial object X together with morphisms $t : X_n \to X_n$ for all $n \geq 0$ such that

$$d_0 t = t d_1; \ d_1 t = t d_2; \ \ldots; d_{n-1} t = t d_n; \ d_n t = d_0 \tag{9}$$

as morphisms $X_n \to X_{n-1}$;
$$s_0 t = t s_1; \; s_1 t = t s_2; \; \ldots; s_{n-1} t = t s_n; \; s_n t = t^2 s_0 \tag{10}$$
as morphisms $X_n \to X_{n+1}$;
$$t^{n+1} = \text{id} \tag{11}$$
on X_n. For a graded algebra A put
$$t(a_0 \otimes a_1 \otimes \ldots \otimes a_n) = (-1)^{|a_0| \sum_{i \geq 1} a_i}(a_1 \otimes a_2 \otimes \ldots \otimes a_0)$$

Proposition 5.2.1 *The above formula for t, together with (6), (7), (8), makes $A^{\#}$ a cyclic graded vector space.*

For a cyclic vector space X define
$$b: X_n \to X_{n-1}; \; b = \sum_{i=0}^{n}(-1)^i d_i; \; \tau = (-1)^{n+1} t: X_n \to X_n; \tag{12}$$
$$N = \sum_{i=0}^{n} \tau^i; \; B: X_n \to X_{n+1}; \; B = (1-\tau)sN \tag{13}$$
where
$$s = t^{-1} s_n : X_n \to X_{n+1}$$

Proposition 5.2.2
$$b^2 = bB + Bb = B^2 = 0$$

Definition 5.4. *For a cyclic vector space put*
$$C_\bullet(X) = (X_\bullet, b)$$
$$CC_\bullet^-(X) = (X_\bullet[[u]], b + uB)$$
$$CC_\bullet(X) = (X_\bullet[u^{-1}, u]]/uX_\bullet[[u]], b + uB)$$
$$CC_\bullet^{\text{per}}(X) = (X_\bullet[u^{-1}, u]], b + uB)$$
where u is a formal parameter of degree -2.

Let C be a small category and $X: C^{op} \to$ k-mod a functor. Let $k^{\#}: C \to$ k-mod be the constant functor whose value on all objects is k and on all morphisms id_k. One defines
$$X \otimes_C k^{\#} = \oplus_{i \in Ob(C)} X(i)/ < a - X(f)a | a \in X(i) \; ; \; f \in \hom_C(j, i) >$$

One can show that $X \mapsto X \otimes_C k^{\#}$ is a right exact functor. Denote its left derived functors by $\text{Tor}_\bullet^C(X, k^{\#})$.

Theorem 5.2.3 *(Connes) One has, for an associative algebra A,*
$$HH_\bullet(A) = \text{Tor}_\bullet^\Delta(A^{\#}, k^{\#})$$
$$HC_\bullet(A) = \text{Tor}_\bullet^\Lambda(A^{\#}, k^{\#})$$

6 Examples

6.1 Smooth functions

For a smooth manifold M one can compute the Hochschild and cyclic homology of the algebra $C^\infty(M)$ where the tensor product in the definition of the Hochschild complex is one of the following three:

$$C^\infty(M)^{\otimes n} = C^\infty(M^n); \qquad (1)$$

$$C^\infty(M)^{\otimes n} = \text{germs}_\Delta\, C^\infty(M^n); \qquad (2)$$

$$C^\infty(M)^{\otimes n} = \text{jets}_\Delta\, C^\infty(M^n) \qquad (3)$$

where Δ is the diagonal.

Theorem 6.1.1 *The map*

$$\mu : f_0 \otimes f_1 \otimes \ldots \otimes f_n \mapsto \frac{1}{n!} f_0 df_1 \ldots df_n$$

defines a quasi-isomorphism of complexes

$$C_\bullet(C^\infty(M)) \to (\Omega^\bullet(M), 0)$$

and a $\mathbb{C}[[u]]$-linear, (u)-adically continuous quasi-isomorphism

$$CC_\bullet^-(C^\infty(M)) \to (\Omega^\bullet(M)[[u]], ud)$$

Localizing with respect to u, we also get quasi-isomorphisms

$$CC_\bullet(C^\infty(M)) \to (\Omega^\bullet(M)[u^{-1}, u]]/u\Omega^\bullet(M)[[u]], ud)$$

$$CC_\bullet^{\text{per}}(C^\infty(M)) \to (\Omega^\bullet(M)[u^{-1}, u]], ud)$$

This theorem, for the tensor products (2, 3), is due essentially to Hochschild, Kostant and Rosenberg (the Hochschild case) and to Connes (the cyclic cases). For the tensor product (1), see [70].

Holomorphic functions

Let M be a complex manifold with the structure sheaf $\mathcal{O}_M$ and the sheaf of holomorphic forms $\Omega_M^\bullet$. If one uses one of the following definitions of the tensor product, then $C_\bullet(\mathcal{O}_M)$, etc. are complexes of sheaves:

$$\mathcal{O}_M^{\otimes n} = \text{germs}_\Delta\, \mathcal{O}_{M^n}; \qquad (4)$$

$$\mathcal{O}_M^{\otimes n} = \text{jets}_\Delta\, \mathcal{O}_{M^n} \qquad (5)$$

where Δ is the diagonal.

Theorem 6.1.2 *The map*

$$\mu : f_0 \otimes f_1 \otimes \ldots \otimes f_n \mapsto \frac{1}{n!} f_0 df_1 \ldots df_n$$

defines a quasi-isomorphism of complexes of sheaves

$$C_\bullet(\mathcal{O}_M) \to (\Omega_M^\bullet, 0)$$

and a $\mathbb{C}[[u]]$-linear, (u)-adically quasi-isomorphism of complexes of sheaves

$$CC_\bullet^-(\mathcal{O}_M) \to (\Omega_M^\bullet[[u]], ud)$$

Similarly for the complexes $CC_\bullet$ and CC^{per}.

6.2 Commutative algebras

Smooth Noetherian algebras

By definition, a commutative Noetherian k-algebra A is smooth if, for any k-algebra C and any ideal I of C such that $I^2 = 0$, the map $\mathrm{Hom}(A, C) \to \mathrm{Hom}(A, C/I)$ is surjective. The class of smooth algebras includes the class of coordinate rings of nonsingular affine varieties over k. Let $\Omega^1_{A/k}$ be the module of Kähler differentials of A and $\Omega^n_{A/k} = \wedge^n_A \Omega^1_{A/k}$.

Theorem 6.2.1 *The map*

$$\mu : a_0 \otimes \ldots a_n \mapsto \frac{1}{n!} a_0 da_1 \ldots da_n$$

defines quasi-isomorphisms of complexes

$$C_\bullet(A) \to (\Omega_{A/k}^\bullet, 0)$$
$$CC_\bullet^-(A) \to (\Omega_{A/k}^\bullet[[u]], ud)$$
$$CC_\bullet(A) \to (\Omega_{A/k}^\bullet[u^{-1}, u]]/u\Omega_{A/k}^\bullet[[u]], ud)$$
$$CC_\bullet^{\mathrm{per}}(A) \to (\Omega_{A/k}^\bullet[u^{-1}, u]], ud)$$

Cf. [44, 45].

Finitely generated commutative algebras

For a finitely generated commutative algebra A over k, choose a surjective homomorphism $f : P \to A$ where P is a ring of polynomials. Let I be the kernel of f. Consider the complexes

$$0 \to P/I^{n+1} \xrightarrow{d} \Omega_{P/k}^1/I^n \xrightarrow{d} \ldots \xrightarrow{d} \Omega_{P/k}^n/I \xrightarrow{d} 0 \qquad (6)$$

for all $n \geq 0$. Denote their cohomologies by $H^*_{\mathrm{cris}}(A/k; n)$ Denote by $H^\bullet_{\mathrm{cris}}(P/k)$ the cohomology of the complex $(\widehat{\Omega}_{P/k}^\bullet, d)$, the I-adic completion of the De Rham complex of P. It is well known that the cohomologies above are independent of a choice of P and f.

Theorem 6.2.2 *1. There exists the canonical morphism*

$$\mu : HC_n(A) \to \bigoplus_{i \geq 0} H^{n-2i}_{cris}(A/k; n-i);$$

if A is smooth then μ is induced by the map from Theorem 6.2.1.
2. If A is a locally complete intersection then μ is an isomorphism.
3. There exists the canonical isomorphism

$$\mu : HC^{per}_\bullet(A) \to H^\bullet_{cris}(A/k)$$

Cf. [27].

6.3 Rings of differential operators

For a C^∞ manifold X of dimension n let $D(X)$ be the ring of differential operators on X. We use the tensor products defined analogously to (2), (3).

Theorem 6.3.1 *([9, 76]). There is a quasi-isomorphism*

$$C_\bullet(D(X)) \to (\Omega^{2n-\bullet}(T^*X), d)$$

which extends to a $\mathbb{C}[[u]]$-linear, (u)-adically continuous quasi-isomorphism

$$CC^-_\bullet(D(X)) \to (\Omega^{2n-\bullet}(T^*X)[[u]], d)$$

As in 6.1, one also has analogous statements for the cyclic and periodic cyclic complexes.

Holomorphic differential operators

Let X be a complex manifold of complex dimension n. For the sheaf D_X of holomorphic differential operators, define the Hochschild, cyclic, etc. complexes of sheaves using tensor products analogous to those in 6.1. Let $\pi : T^*X \to X$ be the projection.

Theorem 6.3.2 *[7] There exists an isomorphism*

$$\pi^{-1} C_\bullet(D_X) \to (\Omega^\bullet_{T^*X}[2n], d)$$

in the derived category of the category of sheaves on T^*X, which extends to a $\mathbb{C}[[u]]$-linear, (u)-adically continuous isomorphism in the derived category

$$\pi^{-1} CC^-_\bullet(D_X) \to (\Omega^\bullet_{T^*X}[2n][[u]], d)$$

As in 6.1, similar isomorphisms exist for the cyclic and periodic cyclic complexes.

6.4 Rings of complete symbols

For a compact smooth manifold X, let $CL(X)$ be the algebra of classical pseudo-differential operators. By $L_\infty(X)$ denote the algebra of smoothing operators (i.e. integral operators with smooth kernel), and put

$$CS(X) = CL(X)/L_\infty(X).$$

We use the projective tensor products.

For any manifold M, denote by $\widehat{\Omega}^*(M \times S^1)$ the space of power series

$$\sum_{\epsilon=0,\,1;\,i=-\infty}^{N} \alpha_{i,\,\epsilon} z^i dz^\epsilon$$

where α_i are forms on M. Denote by S^*X the cosphere bundle of X.

Theorem 6.4.1 *[77] There exists a quasi-isomorphism*

$$C_\bullet(CS(X)) \to (\widehat{\Omega}^{2n-\bullet}(S^*X \times S^1), d)$$

which extends to a $\mathbb{C}[[u]]$-linear, (u)-adically continuous quasi-isomorphism

$$CC_\bullet^-(CS(X)) \to (\widehat{\Omega}^{2n-\bullet}(S^*X \times S^1)[[u]], d)$$

Similarly for $CC_\bullet$, $CC_\bullet^{\text{per}}$. In particular:

Corollary 6.4.2

$$HH_p(CS(X)) = H^{2n-p}(S^*X \times S^1)$$

$$HC_p(CS(X)) = \bigoplus_{i \geq 0} H^{2n-p+2i}(S^*X \times S^1)$$

Combined with the pairing with the fundamental class of $S^*X \times S^1$, the first of the above isomorphisms gives an isomorphism

$$HH_0(CS(X)) = CS(X)/[CS(X), CS(X)] \to \mathbb{C} \qquad (7)$$

This isomorphism is given by the Wodzicki-Guillemin residue [75]. The above theorem has also a holomorphic version where the ring of complete symbols is replaced by the sheaf of microdifferential operators [5].

6.5 Rings of pseudodifferential operators

Theorem 6.5.1

$$HH_0(L_\infty(X)) \simeq \mathbb{C}; \quad HH_p(L_\infty(X)) = 0, \; p > 0;$$

$$HC_{2p}(L_\infty(X)) \simeq \mathbb{C}; \quad HC_{2p+1}(L_\infty(X)) = 0$$

Theorem 6.5.2 *[77]*
$$HH_p(CL(X)) \simeq HH_p(CS(X)) \text{ for } p \neq 1;$$
there is an exact sequence
$$0 \to HH_1(CL(X)) \to HH_1(CS(X)) \to \mathbb{C} \to 0$$

Theorem 6.5.3 *For all $p \geq 0$*
$$HC_{2p}(CL(X)) \simeq HC_{2p}(CS(X))$$
and there is an exact sequence
$$0 \to HC_{2p+1}(CL(X)) \to HC_{2p+1}(CS(X)) \to HC_{2p}(\mathbb{C}) \to 0$$

Theorems 6.5.2 and 6.5.3 follow from Theorem 6.5.1 and from Wodzicki excision theorems [78].

The results of the two previous subsections were extended to more general rings of symbols and of pseudodifferential operators by Melrose-Nistor and Benameur-Nistor ([2, 50]). A survey of these results can be found in [3].

6.6 Noncommutative tori

Let q be a complex number, $|q| = 1$. By A_q denote the algebra over $\mathbb{C}$ with two generators $\mathbf{u}$ and $\mathbf{v}$ subject to one relation
$$\mathbf{vu} = q\mathbf{uv} \tag{8}$$

Theorem 6.6.1 *[12] For q not a root of unity*
$$H_\bullet(A_q) \simeq H_\bullet(\mathbb{T}^2)$$
$$HC_\bullet(A_q) \simeq H_\bullet(\mathbb{T})^2[[u, u^{-1}]]/uH_\bullet(\mathbb{T})^2[[u, u^{-1}]]$$
$$HC_\bullet^-(A_q) \simeq H_\bullet(\mathbb{T}^2)[[u]]$$
$$HC_\bullet^{\text{per}}(A_q) \simeq H_\bullet(\mathbb{T}^2)[[u, u^{-1}]]$$
where u is a formal variable of degree -2.

6.7 Deformation quantization

Let M be a smooth manifold. By a deformation quantization of M we mean a formal product
$$f * g = fg + \sum_{k=1}^{\infty} t^k P_k(f, g) \tag{9}$$
where P_k are bidifferential expressions, $*$ is associative, and $1 * f = f * 1 = f$ for all f. Given such a product (which is called a star product), we define
$$\mathbb{A}^t(M) = (C^\infty(M)[[t]], *) \tag{10}$$

This is an associative algebra over $\mathbb{C}[[t]]$. By $\mathbb{A}_c^t(M)$ we denote the ideal $C_c^\infty(M)[[t]]$ of this algebra. An isomorphism of two deformations is by definition a power series $T(f) = f + \sum_{k=1}^\infty T_k(f)$ where all T_k are differential operators and which is an isomorphism of algebras.

Given a star product on M, for $f, g \in C^\infty(M)$ let

$$\{f, g\} = P_1(f, g) - P_1(g, f) = \frac{1}{t}[f, g]|t = 0 \tag{11}$$

This is a Poisson bracket corresponding to some Poisson structure on M. If this Poisson structure is defined by a symplectic form ω, we say that $\mathbb{A}^t(M)$ is a deformation of the symplectic manifold (M, ω).

Recall the following classification result from [19, 23, 30, 55].

Theorem 6.7.1 *Isomorphism classes of deformation quantizations of a symplectic manifold (M, ω) are in a one-to-one correspondence with the set*

$$\frac{1}{t}[\omega] + H^2(M, \mathbb{C}[[t]])$$

where $[\omega]$ is the cohomology class of the symplectic structure ω.

In defining the Hochschild and cyclic complexes, we use $k = \mathbb{C}[[t]]$ as the ring of scalars, and put

$$\mathbb{A}^t(M)^{\otimes n} = \text{jets}_\Delta C^\infty(M^n)[[t]] \tag{12}$$

Let $\mathbb{A}^t(M)$ be a deformation of a symplectic manifold (M, ω).

Theorem 6.7.2 *There exists a quasi-isomorphism*

$$C_\bullet(\mathbb{A}^t(M), \mathbb{A}^t(M))[t^{-1}] \to (\Omega^{2n-\bullet}(M)[[t, t^{-1}], td)$$

which extends to a $\mathbb{C}[[t, u]]$-linear, (t, u)-adically continuous quasi-isomorphism

$$CC_\bullet^-(\mathbb{A}^t(M))[t^{-1}] \to (\Omega^{2n-\bullet}(M)[[u, t]][t^{-1}], td)$$

An analogous theorem holds for $\mathbb{A}_c^t(M)$ if we replace $\Omega^\bullet$ by $\Omega_c^\bullet$.

The canonical trace

Combining the first map from Theorem 6.7.2 in the compactly supported case with integrating over M and dividing by $\frac{1}{n!}$, one gets the canonical trace of Fedosov

$$Tr : \mathbb{A}_c^t(M) \to \mathbb{C}[[t, t^{-1}]$$

It follows from Theorem 6.7.2 that, for M connected, this trace is unique up to multiplication by an element of $\mathbb{C}[[t, t^{-1}]$.

Deformation quantization of complex manifolds

For a complex manifold M with an open cover $\mathfrak{U}$, define a deformation of the structure sheaf of algebras $\mathcal{O}_M$ to be the following:

a) a collection of holomorphic star products $*_U$ on every $U \in \mathfrak{U}$

b) a collection of holomorphic isomorphisms G_{UV} between $*_U$ and $*_V$ for every $U, V \in \mathfrak{U}$, such that

$$G_{UV}G_{VW} = G_{UW} \tag{13}$$

for all U, V, W.

Such a data defines a sheaf of algebras $\mathbb{A}_M^t$.

(The above definition is enough for the purposes of this article; there is, however, a better definition in which (13) is replaced by $G_{UV}G_{VW} = \text{Ad}(c_{UVW})G_{UW}$ for some invertible elements c_{UVW} subject to a natural cocycle condition on intersections of any four open subsets). Using the definition of tensor products analogous to (1), we define complexes of sheafs $C_\bullet(\mathbb{A}_M^t)$, etc.

Theorem 6.7.3 *Let $\mathbb{A}_M^t$ be a deformation of a complex manifold M with a holomorphic symplectic structure ω. There exists an isomorphism in the derived category of sheaves*

$$C_\bullet(\mathbb{A}_M^t)[t^{-1}] \to (\Omega_M^{2n-\bullet}[[t, t^{-1}], td) \tag{14}$$

which extends to a $\mathbb{C}[[t, u]]$-linear, (t, u)-adically continuous isomorphism in the derived category

$$CC_\bullet^-(\mathbb{A}_M^t)[t^{-1}] \to (\Omega_M^{2n-\bullet}[[t, u]][t^{-1}], td) \tag{15}$$

Similarly for $CC_\bullet$, $CC_\bullet^{\text{per}}$.

If one wants to avoid the language of derived categories, one can say that there may be no morphism of sheaves (15) but there is, canonical up to homotopy, morphism of Čech complexes of corresponding sheaves. One can also use sheaf resolutions other than the Čech resolution, for example the complex of Dolbeault forms with values in Hochschild, cyclic, De Rham, etc. complexes of the algebra of jets [4].

Remark 6.1. The question of classifying deformations of a holomorphic symplectic structure up to isomorphism is more complicated than in the C^∞ case. One can still construct a characteristic class

$$\theta \in \frac{1}{t}[\omega] + H^2(M, \mathbb{C}[[t]])$$

of such a deformation ([4, 54]), but this class does not classify deformations up to isomorphism. See [54] for related results.

6.8 The Brylinski spectral sequence

Consider a deformation quantization $\mathbb{A}^t(M)$ of a smooth manifold M. As we saw above, such a deformation gives rise to a Poisson bracket on $C^\infty(M)$. Denote by π_0 the corresponding Poisson bivector: $\pi_0 \in \Gamma(M, \wedge^2 TM)$ such that $\{f, g\} = <\pi_0, df \wedge dg>$. Denote by i_{π_0} the operator of contracting a form by π_0, and put

$$L_{\pi_0} = di_{\pi_0} - i_{\pi_0} d \tag{16}$$

The Hochschild and cyclic complexes of the deformed algebra $\mathbb{A}^t(M)$ are complete with respect to the filtration $F_p C_\bullet = t^p C_\bullet$. Computing the spectral sequences corresponding to this filtration, one obtains the following

Theorem 6.8.1 *1. There exists a spectral sequence converging to*

$$HH_\bullet(\mathbb{A}^t(M))$$

with the E^1 term equal to the homology of the complex

$$(\Omega^\bullet(M)[[t]], tL_{\pi_0}) \tag{17}$$

2. There exists a spectral sequence converging to

$$HC_\bullet^-(\mathbb{A}^t(M))$$

with the E^1 term equal to the homology of the complex

$$(\Omega^\bullet(M)[[t, u]], tL_{\pi_0} + ud) \tag{18}$$

and similarly for the cyclic and periodic cyclic homologies.

If π_0 is symplectic then the complex (17), resp. (18), is quasi-isomorphic to the complex $(\Omega^{2n-\bullet}(M)[[t]], td)$, resp. $(\Omega^{2n-\bullet}(M)[[t, u]], td)$.

6.9 Deformation quantization: general case

Recall first Kontsevich's classification of deformation quantizations. We call a series

$$\pi = \sum_{i \geq 0} t^i \pi_i,$$

$\pi_i \in \Gamma(M, \wedge^2 T_M)$, *a formal Poisson structure* if $[\pi, \pi] = 0$ where $[\,,\,]$ is the Schouten bracket; in other words, π is a formal Poisson structure if

$$\{f, g\}_\pi = <\pi, df \wedge dg>$$

defines a Lie algebra structure on $C^\infty(M)[[t]]$. Two formal Poisson structures are called *equivalent* if there exists a series

$$X = \sum_{i \geq 0} t^i X_i,$$

$X_i \in \Gamma(M, T_M)$, such that
$$\pi' = \exp \operatorname{ad} L_X(P)$$

Theorem 6.9.1 *(Kontsevich) There is a bijection between the set of isomorphism classes of deformation quantizations of $C^\infty(M)$ and the set of equivalence classes of formal Poisson structures on M.*

For a formal Poisson structure π the above theorem defines up to isomorphism a deformed associative algebra $\mathbb{A}_\pi^t(M)$.

Theorem 6.9.2 *There are quasi-isomorphisms of complexes*
$$C_\bullet(\mathbb{A}_\pi^t(M)) \to (\Omega^\bullet(M)[[t]], tL_\pi)$$
$$CC_\bullet^-(\mathbb{A}_\pi^t(M)) \to (\Omega^\bullet(M)[[t,u]], tL_\pi + ud)$$

In particular, if π is an ordinary Poisson structure, then the Brylinski spectral sequence degenerates.

6.10 Group rings

Let G be a discrete group.

Theorem 6.10.1
$$HH_\bullet(k[G]) \simeq \bigoplus_{<x>} H_\bullet(G_x, k)$$
where the sum is taken over all conjugacy classes of G and G_x is the centralizer of an element x of G.

The Hochschild homology can also be decomposed into components corresponding to conjugacy classes; the construction of these components depends on whether a class is elliptic (i.e. of the form $<x>$ where x is of finite order) or not.

Theorem 6.10.2 *[10]*
$$HC_n(k[G]) \simeq \bigoplus_{<x>|\operatorname{ord} x < \infty} \bigoplus_i H_{n-2i}(G_x, k) \oplus \bigoplus_{<x>|\operatorname{ord} x = \infty} H_{n-2i}(G_x/(x), k)$$

Remark 6.2. The exact sequence of the triangle (5) splits into a direct sum over the set of conjugacy classes of G. For the summands corresponding to elliptic classes, the component of the differential $HH_\bullet \to HC_\bullet$ is injective. In the non-elliptic case, the corresponding component of the exact sequence is isomorphic to the exact sequence of the fibration
$$B\mathbb{Z} \to BG_x \to B(G_x/(x))$$

Simirarly, the components of the spectral sequence (6) which correspond to elliptic classes degenerate at E^1. The component corresponding to a non-elliptic class $<x>$ is isomorphic to the spectral sequence of the fibration
$$BG_x \to B(G_x/(x)) \to K(\mathbb{Z}, 2)$$

7 Index theorems

7.1 Index theorem for deformations of symplectic structures

Let M be a smooth symplectic manifold. Let $\mathbb{A}_M^t$ be a deformation of the sheaf of functions on M. Recall that there exists canonical up to homotopy equivalence quasi-isomorphism (15)

$$\mu^t : CC_\bullet^-(\mathbb{A}^t(M))[t^{-1}] \to (\Omega^{2n-\bullet}(M)[[t,u]][t^{-1}], td) \qquad (1)$$

Localizing in u, we obtain a quasi-isomorphism

$$\mu^t : CC_\bullet^{\mathrm{per}}(\mathbb{A}^t(M))[t^{-1}] \to (\Omega^{2n-\bullet}(M)[[t,u,t^{-1},u^{-1}], td) \qquad (2)$$

Definition 7.1. *The above morphisms are called the trace density morphisms.*

The index theorem compares the trace density morphism to the principal symbol morphism. To define the latter, consider the cyclic complex of the deformed algebra where the scalar ring is $\mathbb{C}$ instead of $\mathbb{C}[[t]]$. Consider the composition

$$CC_\bullet^-(\mathbb{A}^t(M)) \to CC_\bullet^-(C^\infty(M)) \to$$
$$\to (\Omega^\bullet(M)[[u]], ud) \to (\Omega^\bullet(M)[[u]][[t]], ud)$$

where the first morphism is reduction modulo t, the second one is μ from Theorem 6.1.1, and the third one is induced by the embedding $\mathbb{C} \to \mathbb{C}[[t]]$. We will denote this composition, followed by localization in t, by

$$\mu : CC_\bullet^{\mathrm{per}}(\mathbb{A}^t(M)) \to (\Omega^\bullet(M)[[u,t,u^{-1},t^{-1}], ud) \qquad (3)$$

To compare μ and μ^t, let us identify the right hand sides by the isomorphism

$$(\Omega^{2n-\bullet}(M)[[u,t,u^{-1},t^{-1}], td) \to (\Omega^\bullet(M)[[u,t,u^{-1},t^{-1}], ud)$$

which is equal to $(\frac{t}{u})^{n-k}$ on $\Omega^k(M)[[u,t,u^{-1},t^{-1}]$. After this identification, we obtain two morphisms

$$\mu, \mu^t : CC_\bullet^{\mathrm{per}}(\mathbb{A}^t(M)) \to (\Omega^\bullet(M)[[u,t,u^{-1},t^{-1}], ud)$$

where the left hand side is defined as the periodic cyclic complex with respect to the ground ring $\mathbb{C}$.

Theorem 7.1.1 *At the level of cohomology,*

$$\mu^t = \sum_{p=0}^{\infty} u^p (\widehat{A}(M) e^\theta)_{2p} \cdot \mu$$

where $\widehat{A}(M)$ is the $\widehat{A}$ class of the tangent bundle of M viewed as a complex bundle (with an almost complex structure compatible to the symplectic form), and $\theta \in \frac{1}{t}[\omega] + H^2(M, \mathbb{C}[[t]])$ is the characteristic class of the deformation (cf. Theorem 6.7.1).

Note that the canonical trace $\mathrm{Tr}_{\mathrm{can}}$ is the composition of μ^t with the integration $\Omega^{2n}[[t,t^{-1}]] \to \mathbb{C}[[t,t^{-1}]]$. Let P and Q be $N \times N$ matrices over $\mathbb{A}^t(M)$ such that $P^2 = P$, $Q^2 = Q$, and $P - Q$ is compactly supported. Let P_0, Q_0 be reductions of P, Q modulo t. They are idempotent matrix-valued functions; their images $P_0 \mathbb{C}^N$, $Q_0 \mathbb{C}^N$ are vector bundles on M. Applying μ^t to the the difference of Chern characters of P and Q, we obtain the following index theorem of Fedosov [30] (cf. also [55]).

Theorem 7.1.2

$$\mathrm{Tr}_{\mathrm{can}}(P - Q) = \int_M (\mathrm{ch}(P_0 \mathbb{C}^N) - \mathrm{ch}(Q_0 \mathbb{C}^N)) \widehat{A}(M) e^\theta$$

7.2 Index theorem for holomorphic symplectic deformations

Let M be a complex manifold with a holomorphic symplectic form ω. Let $\mathbb{A}^t_M$ be a deformation of the sheaf of holomorphic functions on M. One defines two morphisms in the derived category of sheaves

$$\mu, \mu^t : CC^{\mathrm{per}}_\bullet(\mathbb{A}^t_M) \to (\Omega^\bullet_M[[u, t, u^{-1}, t^{-1}]], ud)$$

exactly as in the smooth case.

Theorem 7.2.1

$$\mu^t = \sum_{p=0}^\infty u^p (\sqrt{\mathrm{Td}(M)} e^\theta)_{2p} \cdot \mu$$

in the derived category of sheaves, where $\mathrm{Td}(M)$ is the Todd class of the holomorphic tangent bundle of M.

7.3 General index theorem for deformations

Let M be a C^∞ manifold. Let π be a formal Poisson structure on M and $\mathbb{A}^t_\pi(M)$ the deformation corresponding to π under the Kontsevich isomorphism from Theorem 6.9.1. We construct the quasi-isomorphisms

$$\mu, \mu^t : CC^{\mathrm{per}}_\bullet(\mathbb{A}^t_\pi(M)) \to (\Omega^\bullet(M)[[u, t, u^{-1}, t^{-1}]], ud)$$

as follows. The principal symbol map μ is constructed exactly as in (3). The trace density morphism μ^t is defined as follows. Start with the quasi-isomorphism

$$CC^-_\bullet(\mathbb{A}^t_\pi(M)) \to (\Omega^\bullet(M)[[t, u]], tL_\pi + ud)$$

from Theorem 6.9.2. Then localize it with respect to t and u, and then follow it by the isomorphism

$$\exp(-\frac{t}{u} i_\pi) : (\Omega^\bullet(M)[[t, u, t^{-1}, u^{-1}]], tL_\pi + ud) \to (\Omega^\bullet(M)[[t, u, t^{-1}, u^{-1}]], ud).$$

Theorem 7.3.1 *At the level of cohomology,*

$$\mu^t = \sum_{p=0}^{\infty} u^p \tau_{2p} \cdot \mu$$

where the De Rham cohomology class $\tau = \sum \tau_{2p}$ *is as in Theorem 4.4.2.*

Theorem 7.1.1 can be deduced from the theorem above. Note the absence of the factor e^θ, explained by the presence of $\exp(-\frac{t}{u}i_\pi)$ in the definition of the trace density map μ^t.

Let us stress again that all the ingredients of the theorem above (the construction of the deformation $\mathbb{A}_\pi^t(M)$) from π and the construction of the trace density map) are not canonical but depend on a choice of a Drinfeld associator.

8 Riemann-Roch theorem for D-modules

Let us recall the definition and examples from [60]. Let X be a complex manifold. An elliptic pair $(\mathcal{M}, \mathcal{F})$ on X is a coherent D_X-module $\mathcal{M}$ and an $\mathbb{R}$-constructible sheaf $\mathcal{F}$ on X such that

$$\text{char}(\mathcal{M}) \bigcap SS(\mathcal{F}) \subset T_X^* X \qquad (1)$$

More generally, $(\mathcal{M}, \mathcal{F})$, can be a pair of complexes, or rather of objects of corresponding derived categories.

Examples

1. Let $\mathcal{M}$ be a coherent D_X-module. Then the pair $(\mathcal{M}, \mathbb{C}_X)$ is elliptic.
2. Let U be an open subset of X. Then the pair $(\mathcal{M}, \mathbb{C}_U)$ is elliptic if and only if ∂U is noncharacteristic for $\mathcal{M}$.
3. Let X be a complexification of a real analytic manifold X_0 and $\mathcal{M}$ a complexification of a real analytic D_{X_0}-module $\mathcal{M}_0$ on X_0. Then $(\mathcal{M}, \mathbb{C}_{X_0})$ is an elliptic pair if and only if $\mathcal{M}_0$ is elliptic in the classical sense.
4. If $\mathcal{F}$ is $\mathbb{R}$-constructible on X then $(\mathcal{O}_X, \mathcal{F})$ is an elliptic pair.

We assume that $\mathcal{M}$ is endowed with a good filtration. Recall that the order of differential operators defines a filtration F_p on D_X; a filtration G_q on a module is good if $F_p D_X \cdot G_q \subset G_{p+q}$.

Consider the derived tensor product $\mathcal{M} \otimes^{\mathbb{L}}_{D_X} \mathcal{O}_X$. Schapira and Schneiders showed that $(\mathcal{M} \otimes^{\mathbb{L}}_{D_X} \mathcal{O}_X) \otimes_\mathbb{C} \mathcal{F}$ has finite-dimensional cohomology. Let $\mu\, \text{eu}(\mathcal{F})$ be Kashiwara's microlocal Euler class of $\mathcal{F}$. Let Λ be the characteristic variety of $\mathcal{M}$. The good filtration on $\mathcal{M}$ defines a coherent $\text{gr}_F D_X$-module $\text{gr}\,\mathcal{M}$; since $\text{gr}_F D_X$ is the sheaf of functions on T^*X which are polynomial along the fibres, by extension of scalars we obtain a coherent $\mathcal{O}_{T^*X}$-module $\sigma_\Lambda(\mathcal{M})$ supported on Λ. One can define Chern character ch_Λ of such a module in $H^\bullet_\Lambda(T^*X)$.

The following was conjectured in [60] and proven in [4] by reducing it to Theorem 7.2.1.

Theorem 8.0.2

$$\chi(R\Gamma((\mathcal{M} \otimes^{\mathbb{L}}_{\mathcal{D}_X} \mathcal{O}_X) \otimes_{\mathbb{C}} \mathcal{F})) = \int \mu \operatorname{eu}(\mathcal{F}) \smile \operatorname{ch}_\Lambda(\sigma_\Lambda(\mathcal{M})) \smile \pi^* \operatorname{Td}(X)$$

where π is the projection from T^*X to X.

References

1. F. Bayen, M. Flato, C. Fronsdal, A. Lichnerowicz, D. Sternheimer, *Deformation theory and quantization*, Ann. Phys. **111** (1977), 61–151.
2. M. Benameur, V. Nistor, *Homology of complete symbols and noncommutative geometry*, QA/0008131.
3. M. Benameur, J. Brodzki, V. Nistor, *Cyclic homology and pseudodifferential operators*, Quantization of singular symplectic quotients, 21–46, Progr. Math., **198**, Birkhuser, Basel, 2001.
4. P. Bressler, R. Nest, B. Tsygan, *Riemann-Roch theorems via deformation quantization I and II*, Advances in Mathematics **67**(1) (2002), 1–73.
5. P. Bressler, R. Nest, B. Tsygan, *A Riemann-Roch type formula for the microlocal Euler Class*, Int. Math. Res. Notices **20** (1997), 1033–1044.
6. J. Brodzki, *Semi-direct products, derivations and reduced cyclic homology*, J. reine angew. Math. **436** (1993), 177–195.
7. J.-L. Brylinski, *Some examples of Hochschild and cyclic homology*, Utrecht 1986, A. Cohen et al. (eds.), Lecture Notes in Math. vol. 1271, Springer-Verlag 1987, pp. 33–72.
8. J.-L. Brylinski, *Differential complex of a Poisson manifold*, Journal of Differential Geometry.
9. J.-L. Brylinski, E. Getzler, *The homology of algebras of pseudodifferential operators and the noncommutative residue*, K-theory **1**(4) (1987), 385–403.
10. D. Burghelea, *The cyclic homology of group rings*, Comment. Math. Helv. **60** (1985), 354–365.
11. H. Cartan, S. Eilenberg, *Homological Algebra*, Princeton, 1956.
12. A. Connes, *Noncommutative Geometry*, New York-London, Academic Press, 1994.
13. A. Connes, *noncommutative differential geometry*, IHES Publ. Math. **62** (1985), 257–360.
14. A. Connes, M. Flato, D. Sternheimer, *Closed star product and cyclic cohomology*, Lett. Math. Phys. **24**(1) (1992), 1–12
15. M. Chas, D. Sullivan, *String topology*, Math.GT9911159.
16. A. Connes, H. Moscovici, *Cyclic homology and Hopf algebras*, Lett. Math. Phys **48** (1999), 97–108.
17. A. Connes, H. Moscovici, *Hopf algebras, cyclic homology and the transverse index theorem*, CMP **198** (1998), 199–246.
18. A. Connes, H. Moscovici, *The local index formula in noncommutative geometry*, GAFA **5** (1995), 174–243.

19. P. Deligne, *Déformations de l'algèbre des fonctions d'une variété symplectique, Comparaison entre Fedosov et Dewilde, Lecomte*, Selecta Math., New series, **1** (1995), 667–698.
20. Yu. Daletski, B. Tsygan, *Operations on cyclic and Hochschild complexes*, Methods Funct. Anal. Topology **5** (1999), 4, 62–86.
21. Yu. Daletski, I. Gelfand and B. Tsygan, *On a variant of noncommutative geometry*, Soviet Math. Dokl. **40** (1990), 2, 422–426.
22. V. Drinfeld, *On quasi-triangular Hopf algebras and on a group that is closely connected with* $\text{Gal}(\overline{\mathbb{Q}})/\mathbb{Q}$, Leningrad J. Math. **2** (1991), 4, 829–860.
23. De Wilde, P. Lecomte, *Existence of star products and of formal deformations of the Poisson Lie algebra of arbitrary symplectic manifolds*, Lett. Math. Phys. **7** (1983), 487–496.
24. P. Etingof and D. Kazhdan, *Quantization of Lie bialgebras, II*, Selecta Math. **4** (1998), 213–231
25. C. Epstein, R. Melrose, *Contact degree and the index of Fourier integral operators*, Matematics Research Letters**5**(3) (1998), 363–381.
26. B. Feigin and B. Tsygan, *Additive K-theory*, LMN 1289 (1987), 66–220.
27. B. Feigin and B. Tsygan, *Cohomology of Lie algebras of generalized Jacobi matrices*, Funct. Anal. and Appl. **17** (1983).
28. B. Feigin and B. Tsygan, *Lie algebra homology and the Riemann-Roch theorem*, Proccedings of the winter school of 2nd winter school at Srni, Rend. Math. Palermo (1989).
29. B. Feigin and B. Tsygan, *Additive K-theory and cristalline cohomology*, Funct. Anal. and Appl. **19** (1985).
30. B. Fedosov, *Deformation Quantization and Index Theorem*, Akademie Verlag, 1994.
31. K. Fukaya, *Floer homology of Lagrangian foliation and noncommutative mirror symmetry*, Kyoto University preprint 98-08.
32. M. Gerstenhaber, *The Cohomology structure of an associative ring*, Ann. Math. **78** (1963), 267–288.
33. M. Gerstenhaber, A. Voronov, *Homotopy G-algebras and moduli space operad*, IMRN (1995), 141–153.
34. E. Getzler, *Batalin-Vilkovisky algebras and two-dimensional topological field theories*, CMP **159** (1994), 265–285.
35. E. Getzler, *Cartan homotopy formulas and the Gauss-Manin connection in cyclic homology*, Israel Math. Conf. Proc. **7**, 65–78.
36. E. Getzler, J. Jones, *Operads, homotopy algebra and iterated integrals for double loop spaces*, hep-th9403055.
37. E. Getzler and J. Jones, A_∞ *algebras and the cyclic bar complex*, Illinois J. of Math. **34** (1990), 256–283.
38. G. Halbout, Thèse doctorale, Université de Strasbourg, 1999.
39. C.E. Hood and J.D.S. Jones, *Some algebraic properties of cyclic homology groups*, K-Theory, **1** (1987), 361–384.
40. M. Kashiwara, P. Schapira, *Sheaves on Manifolds*, Grundlehren der mathematischen Wissenschaften, no. 292, Springer, 1990.
41. M. Kontsevich, *Deformation quantization of Poisson manifolds I*, QA/9709040.
42. M. Kontsevich, *Homological algebra of mirror symmetry*, Proc. ICM I (1994), 120–139.

43. M. Kontsevich, Y. Soibelman, *Homological mirror symmetry and torus fibrations*, Symplectic Geometry and Mirror Symmetry (Seoul, 2000), World Sci. Publishing, 2001.
44. J.-L. Loday, *Cyclic Homology*, Springer Verlag, 1993.
45. J.-L. Loday and D. Quillen, *Cyclic homology and the Lie algebra homology of matrices*, Comment. Math. Helv. **59** (1984), 565–591.
46. E. Leichtnam, R. Nest, and B. Tsygan, *Local index formula for the index of a Fourier integral operator*, Journal of Differential Geometry **59**(2) (2001), 269–300
47. T. Lada, J.D. Stasheff, *Introduction to sh algebras for physicists*, International Journal of Theor. Physics **32** (1993), 1087–1103
48. Yu. Manin, *Topics in Noncommutative Geometry*, M.B. Porter Lectures, Princeton University Press, Princeton, NJ, 1991.
49. K. MacKenzie, *Lie Groupoids and Lie Algebroids in Differential Geometry*, Lecture Notes Series 124, London Mathematical Society, (1987).
50. R. Melrose, V. Nistor, *Homology of pseudodifferential operators I. Manifolds with boundary*, funct-an/9606005.
51. R. Nest, B. Tsygan, *On the cohomology ring of an algebra*, QA/9803132, Advances in Geometry, in: Progress in Mathematics, Birkhäuser, **172** (1997).
52. R. Nest, B. Tsygan, *Algebraic Iindex theorem for families*, Adv.Math. **113** (1995), 2, 151–205.
53. R. Nest, B. Tsygan, *Formal versus analytic index theorems*, IMRN **11**, 1996, 557–564.
54. R. Nest, B. Tsygan, *Deformations of symplectic Lie algebroids, deformations of holomorphic symplectic structures, and index theorems*, Asian J. Math. **5**(4) (2001), 599–635.
55. R. Nest, B. Tsygan, *Algebraic index theorem*, Com. Math. Phys **172** (1995), 2, 223–262.
56. R. Nest, B. Tsygan, *Deformations of manifolds with boundary*, Krelle J. für Reine und Ang. Math. (1996).
57. R. Nest, B. Tsygan, *Fukaya type categories for associative algebras*, Deformation Theory and Symplectic Geometry, Mathematical Physics Studies, vol. 20, Kluwer Acad. Publ. (1997).
58. G. Rinehart, *Differential forms on general commutative algebras*, Trans. AMS **108** (1963), 139–174.
59. Y. Soibelman, *Quantum tori, mirror symmetries, and deformation theory*, Lett. Math. Phys. **56** (2001), 99–125.
60. P. Schapira and J.-P. Schneiders, *Index Theorem for Elliptic Pairs*, II, Astérisque vol. 224, 1994.
61. A. Schwarz, *Geometry of Batalin-Vilkovisky quantization* , CMP **158** (1993), 373–396.
62. J. Stasheff, *Closed string theory, strong homotopy Lie algebras and the operad actions of moduli space*, hep-th/9304061.
63. J. Stasheff, *Homotopy associativity of H-spaces*, Trans. AMS **108** (1963), 275–292.
64. D. Tamarkin, *Another proof of M. Kontsevich formality theorem*, QA/9803025.
65. D. Tamarkin, *Formality of chain operads of small squares*, QA/9809164.
66. D. Tamarkin, B. Tsygan, *Cyclic formality and index theorems*, Letters in Mathematical Physics **56**(2) (2001), 85–97.

67. D. Tamarkin, B. Tsygan, *Noncommutative differential calculus, homotopy BV algebras and formality conjectures*, Methods of Functional Analysis and topology, 1, 2001.
68. D. Tamarkin, B. Tsygan, *Formality conjectures for chains*, AMS Translations **194**(2) (2000), 261–276.
69. D. Tamarkin, B. Tsygan, *Differential operators on differential forms in noncommutative differential calculus*, to appear.
70. N. Teleman, *Microlocalisation de l'homologie de Hochschild*, C. R. Acad. Sci. Paris Sr. I Math. **326**(11) (1998), 1261–1264.
71. B. Tsygan *Homology of Lie algebras of matrices over rings and Hochschild homology*, Uspekhi Mat. Nauk **38**(2) (1983), 217–218.
72. B. Tsygan *Formality conjectures for chains*, Amer. Math. Soc. Transl., **194**(2) (1999), 261–274.
73. A. Voronov, *Topological field theories, string backgrounds and homotopy algebras*, Proc. of XXII Conf. on Diff. Geom. Methods in Theor. Phys., **4** (1994), 161–178.
74. A. Weinstein, *Deformation quantization*, Séminaire Bourbaki, Vol. 1993/94, Astérisque No. 227 (1995), Exp. 789, 5, 389–409.
75. M. Wodzicki, *Noncommutative residue, Part I: fundamentals*, Lecture notes in Math. **1289** (1987).
76. M. Wodzicki, *Cyclic homology of differential operators*, Duke Math. J. **54**(2) (1987), 641–647.
77. M. Wodzicki, *Cyclic homology of pseudodifferential operators and noncommutative Euler class*, C.R.A.S. Paris **306**(6) (1988), 321–325.
78. M. Wodzicki, *Excision in cyclic homology and in rational algebraic K-theory*, Ann. Math. **129** (1989), 591–640.

Noncommutative Geometry, the Transverse Signature Operator, and Hopf Algebras
[after A. Connes and H. Moscovici]*

Georges Skandalis (translated by Raphaël Ponge and Nick Wright)

1	**Preliminaries**	117
1.1	Complex powers of pseudodifferential operators; the Wodzicki-Guillemin residue	117
1.2	Cyclic cohomology	117
1.3	K-homology; p-summable cycles	118
2	**The Local Index Formula**	120
3	**The Diff-Invariant Signature Operator**	122
3.1	The Hilbert space $\mathcal{H}$ and the algebra $\mathcal{A}$	123
3.2	The vertical part of the signature operator	124
3.3	The horizontal part of the signature operator	124
3.4	The spectral triple	124
4	**The "Transverse" Hopf Algebra**	125
4.1	Matched pairs of groups	125
4.2	The Hopf algebra $\mathcal{H}_n$	127
4.3	Cyclic cohomology of Hopf algebras	128
4.4	The classifying map	129
4.5	Cyclic cohomology of the Hopf algebra $\mathcal{H}_n$	129
4.6	The cyclic cocycle Φ	130
4.7	Calculation of Φ	131
4.8	A variant of the Hopf algebra	132
	References	132

---

* Originally published as "Géométrie non commutative, opérateur de signature transverse et algèbres de Hopf" in Séminaire Bourbaki, 53ème année, 2000–2001, no. 892

One of the main goals of noncommutative geometry (cf. [13, 27]) is to translate the tools of differential geometry into a purely operator theoretic language; from this point of view a noncommutative manifold is represented by a noncommutative algebra which formally plays the role of the algebra of smooth functions on the ghost noncommutative manifold. The leading example is given by the noncommutative manifold V/F, the space of the leaves of a foliation F on a manifold V: by picking an (open) transversal M we are lead to consider the noncommutative algebra obtained as the crossed product of the algebra $C_c^\infty(M)$ of compactly supported smooth functions on M by a pseudogroup of diffeomorphisms.

Assume then that Γ is a countable group of smooth diffeomorphisms on a manifold M of dimension n. The aim of the work of Connes-Moscovici, surveyed in this article, is to set up an index theorem in this context. The first problem is that in general there exists no Γ-invariant metric on M, so that there cannot exist an elliptic differential operator with a Γ-invariant principal symbol. To overcome this, the authors shift to the total space P of the bundle of metrics, that is the frame space modulo the $O(n)$-action; they then build a pseudodifferential operator D which is hypoelliptic and almost invariant by *all the diffeomorphisms* of M.

The index formula for these operators is given by a cyclic cocycle φ on the crossed product algebra $C_c^\infty(P) \rtimes \Gamma$. We can then give an index formula built out of operator traces, but the latter is not local, and thus cannot be computed in local charts. To remedy this the authors prove an index formula which holds in a much more general operator theoretic setting, and involves higher analogues of the Wodzicki-Guillemin residue (i.e. residues of meromorphic functions of the type $z \mapsto \mathrm{Tr} b|D|^{-z}$).

At this stage, we are far from being done: even in dimension 1 (that is for codimension 1 foliations) the index formula we get has thousands of terms, most of them giving a zero contribution that we must suppress in all dimensions. To this end Connes and Moscovici introduce a noncommutative and noncocommutative Hopf algebra $\mathcal{H}_n$ of *transverse vector fields* on $\mathbb{R}^n$ playing the role of a (quantum) symmetry group. This allows them to *a priori* simplify the computations. Better: they construct a Hopf cyclic cohomology $\mathrm{HC}^*_{\mathrm{Hopf}}(\mathcal{H}_n)$ for $\mathcal{H}_n$ and a universal cocycle $\Phi \in \mathrm{HC}^*_{\mathrm{Hopf}}(\mathcal{H}_n)$, and they show that the index formula (i.e. the cocycle φ) comes from Φ through a *classifying* map

$$\mathrm{HC}^*_{\mathrm{Hopf}}(\mathcal{H}_n) \to \mathrm{HC}^*(C_c^\infty(P) \rtimes \Gamma).$$

Furthermore, they show the cyclic cohomology $\mathrm{HC}^*_{\mathrm{Hopf}}(\mathcal{H}_n)$ is canonically isomorphic to the Gel'fand-Fuchs cohomology $H^*(WO(n))^2$. Finally, they compute the class of Φ in the Gel'fand-Fuchs cohomology.

To let the Hopf algebra $\mathcal{H}_n$ act, we need first to chose a covering of our transversal M by open subsets of $\mathbb{R}^n$; we can actually avoid this step by building another Hopf algebra $\mathcal{H}_M$, which is a quantum groupoid rather than

---

[2] In fact, it is equal to a variant $H^*(WSO(n))$.

a quantum group and allows us to define the classifying map in terms of a (torsion free) affine connection.

We start this survey by recalling some basic facts concerning cyclic cohomology, K-homology and the Wodzicki-Guillemin residue trace. Then, we explain the purely operator theoretic index formula. Next, we present the construction of the transverse signature operator, and more generally of a diffeomorphism invariant pseudodifferential calculus. In the last section, we give the construction of the Hopf algebra $\mathcal{H}_n$ and explain how it enables us to deduce an index formula for the transverse signature operator.

I wish to thank Alain Connes and Henri Moscovici as well as Saad Baaj for their advice and help during the preparation of this survey, and for their careful reading of a preliminary version of this manuscript.

1 Preliminaries

1.1 Complex powers of pseudodifferential operators; the Wodzicki-Guillemin residue (cf. [32, 39, 44])

Let M be a smooth compact manifold with dimension n. Let P be a (classical) elliptic (pseudo)differential operator of integer degree $m > 0$. We assume here that P is one to one and positive (as a L^2 unbounded operator). The operator P^s, $s \in \mathbb{C}$, is the prototype of a pseudodifferential operator with order ms. If Q is a (classical) pseudodifferential operator of integer order k, the function $m^{-1}\mathrm{Tr}(QP^{-s})$, defined for $m\,\mathrm{Re}(s) > k + n$, has a unique meromorphic continuation to $\mathbb{C}$ with at worst simple poles on numbers $k + n - \frac{\ell}{m}$, $\ell \in \mathbb{N}$. The residue $\mathrm{res}Q$ at $s = 0$ does not depend on the choice of P; this is the Wodzicki-Guillemin residue of Q. The latter has a local expression $\mathrm{res}Q = \int_M s_Q$ where s_Q is a density on M computable by means of the (total) symbol of Q: for $x \in M$, $s_Q(x)$ is the integral over the cotangent sphere at x of the symbol of Q of degree $-n$ (in any local chart for M). Further, this residue is a trace: if P, Q are pseudodifferential operators with integer orders, then $\mathrm{res}PQ = \mathrm{res}QP$. If M is connected and has dimension > 1 it is in fact the only trace on the algebra of pseudodifferential operators on M with integer orders modulo the smoothing operators (i.e. those with a smooth Schwartz kernel).

1.2 Cyclic cohomology (cf. [7, 10, 13, 42])

Let $\mathcal{A}$ be an (associative and unital) algebra over $\mathbb{C}$. Let us consider the space of multilinear forms $\mathcal{A}^{n+1} \to \mathbb{C}$ as the space $C^n(\mathcal{A}, \mathcal{A}^*)$ of n-cochains on the $(\mathcal{A}, \mathcal{A})$-bimodule $\mathcal{A}^*$. The Hochschild boundary b is then given by

$$(b\varphi)(a_0, a_1, \ldots, a_n, a_{n+1}) = \sum_{k=0}^{n}(-1)^k \varphi(a_0, \ldots, a_k a_{k+1}, \ldots, a_{n+1})$$
$$+(-1)^{n+1}\varphi(a_{n+1}a_0, a_1, \ldots, a_n).$$

For $\varphi \in C^n(\mathcal{A}, \mathcal{A}^*)$, we let $\varphi^\lambda = \lambda(\varphi) \in C^n(\mathcal{A}, \mathcal{A}^*)$ be the form given by

$$\varphi^\lambda(a_0, a_1 \ldots, a_n) = (-1)^n \varphi(a_1, \ldots, a_n, a_0).$$

We denote by $C_\lambda^n(\mathcal{A})$ the subspace of $C^n(\mathcal{A}, \mathcal{A}^*)$ of *cyclic* cochains, i.e. such that $\varphi^\lambda = \varphi$. It is easy to check that $b(C_\lambda^n(\mathcal{A})) \subset C_\lambda^{n+1}(\mathcal{A})$. The cyclic cohomology of $\mathcal{A}$ is the cohomology of

$$\ldots \xrightarrow{b} C_\lambda^n(\mathcal{A}) \xrightarrow{b} C_\lambda^{n+1}(\mathcal{A}) \xrightarrow{b} \ldots.$$

Let $\varphi \in C^{2n}(\mathcal{A}, \mathcal{A}^*)$ be a cyclic cocycle and let $e \in \mathcal{A}$ be an idempotent. One can check that $\varphi(e, e, \ldots, e)$ depends only on the class φ in the cyclic cohomology $\mathrm{HC}^{2n}(\mathcal{A})$ of $\mathcal{A}$ and on the class of e in the K-theory of $\mathcal{A}$. We thus get a pairing between the cyclic cohomology of $\mathcal{A}$ and its K-theory.

The cyclic cohomology of $\mathcal{A}$ is also the cyclic cohomology of the following bicomplex: $(C_{p,q}, b, B)_{p,q \in \mathbb{N};\ q \leq p}$ where $C_{p,q} = C^{p-q}(\mathcal{A}, \mathcal{A}^*)$ and $B : C^n(\mathcal{A}, \mathcal{A}^*) \to C^{n-1}(\mathcal{A}, \mathcal{A}^*)$ is the map $A \circ t \circ D$ with

$$D = \mathrm{id} - \lambda : C^n(\mathcal{A}, \mathcal{A}^*) \to C^n(\mathcal{A}, \mathcal{A}^*),$$

$$A = \sum_{k=0}^{n-1} \lambda^k : C^{n-1}(\mathcal{A}, \mathcal{A}^*) \to C^{n-1}(\mathcal{A}, \mathcal{A}^*),$$

and $(t\varphi)(a_0, a_1, \ldots, a_{n-1}) = \varphi(1, a_0, a_1, \ldots, a_{n-1})$.

An n-cocycle C in this bicomplex is thus a sequence φ_{n-2j}, $0 \leq j \leq [n/2]$, such that $B\varphi_{n-2j} = -b\varphi_{n-2j-2}$ for any j. If n is even, the pairing of C with an idempotent e is

$$\langle C | e \rangle = \sum_{j=0}^{n/2} (-1)^j \frac{(2j)!}{j!} \varphi_{2j}\left(e - \frac{1}{2}, e, \ldots, e\right).$$

The *periodic cyclic cohomology* $\mathrm{HC}^*_{\mathrm{per}}(\mathcal{A})$ of $\mathcal{A}$ is the cohomology of the bicomplex $(C_{p,q}, b, B)_{p,q \in \mathbb{Z};\ q \leq p}$, which is obviously 2-periodic. The pairing of $\mathrm{HC}^*(\mathcal{A})$ with the K-theory of $\mathcal{A}$ factors through the natural homomorphism $\mathrm{HC}^*(\mathcal{A}) \to \mathrm{HC}^*_{\mathrm{per}}(\mathcal{A})$.

1.3 K-homology; p-summable cycles (cf. [1, 6, 10, 31])

The starting point of noncommutative geometry is K-homology. It was first described by Atiyah [1] in terms of abstract pseudodifferential operators and by Brown-Douglas-Fillmore [6] using C^*-algebra extensions, after which it was thoroughly developed by Kasparov [31].

Recall that a K-cycle of a C^*-algebra A is given by a pair (π_+, π_-) of representations of A on Hilbert spaces $H_\pm$ and a Fredholm operator $T : H_+ \to$

H_- intertwining the representations $\pi_\pm$ up to compact operators, so that $\pi_-(a)T - T\pi_+(a) \in \mathcal{K}(H_+, H_-)$ for any $a \in A$. The cycle $(H_\pm, \pi_\pm, T)$ defines a homomorphism ψ from $K_0(A)$ to $\mathbb{Z}$ as follows. If e is an idempotent in $M_n(A)$, then the operator $T_e = \pi_-(e)(T \otimes 1)\pi_+(e)$ from $\pi_+(e)(H_+ \otimes \mathbb{C}^n)$ to $\pi_-(e)(H_- \otimes \mathbb{C}^n)$ is Fredholm, where the natural representation of $M_n(A) = A \otimes M_n(\mathbb{C})$ on $(H_\pm)^n = H_\pm \otimes \mathbb{C}^n$ is again denoted by $\pi_\pm$; its index depends only on the K-theory class of e and one defines $\psi([e]) = \text{ind}(T_e)$.

We often need to suppose that T is invertible, possibly by changing $H_\pm$ by finite dimensional spaces (and by removing the condition $\pi_\pm(1) = 1$). Then by standard functional calculus we can even suppose T is unitary (i.e. $T^{-1} = T^*$).

It is useful to use $\mathbb{Z}/2\mathbb{Z}$-graded notations: we consider the space $H = H_+ \oplus H_-$, the representation $\pi : a \mapsto \begin{pmatrix} \pi_+(a) & 0 \\ 0 & \pi_-(a) \end{pmatrix}$ and the operator $F = \begin{pmatrix} 0 & T \\ T^{-1} & 0 \end{pmatrix}$. From this point of view, the cycles are then triples (H, π, F), where H is a $\mathbb{Z}/2\mathbb{Z}$-graded Hilbert space, $\pi : A \to \mathcal{L}(H)$ is a representation such that the operator $\pi(a)$ has degree 0 with respect to the grading (i.e. it preserves it) for any $a \in A$, and the operator $F \in \mathcal{L}(H)$ has degree 1 with respect to the grading, $F^2 = 1$ and $[\pi(a), F] \in \mathcal{K}(H)$ for any $a \in A$. Usually we will identify A with its image under the homomorphism π.

A first observation, made by Connes at the beginning of the 80's, is the following: if we assume the idempotent $e \in A$ $[F, e]$ to be in a Schatten class $\mathcal{L}^p(H)$ (i.e. the operator $||[F, e]||^p$ is trace class), then we can write the index of T_e as a trace,

$$\text{ind}(T_e) = (-1)^k \text{Tr}(\gamma e[F, e]^{2k}), \tag{1}$$

where γ is the grading operator, i.e. $\gamma = \begin{pmatrix} 1 & 0 \\ 0 & -1 \end{pmatrix}$, and $k \in \mathbb{N}$ is such that $2k \geq p$. This formula can also be written as $\text{ind}(T_e) = (-1)^k \varphi(e, e, e, \ldots, e)$, where φ is the $2k+1$-linear form on the subalgebra $\mathcal{A} = \{a \in A; [F, a] \in \mathcal{L}^p(H)\}$ of A given by $\varphi(a_0, a_1, \ldots, a_{2k}) = \text{Tr}(\gamma a_0 [F, a_1] \ldots [F, a_{2k}])$. It is easy to see that φ is a cyclic cocycle. Indeed, it was the analysis of the properties of φ used to define a homomorphism from the K-theory of $\mathcal{A}$ into $\mathbb{C}$ that lead Connes to invent cyclic cohomology.

In general, the subalgebra $\mathcal{A}$ does not coincide with A. We say that the cycle (H, F) is p-summable if $\mathcal{A}$ is dense in A (in that case $\mathcal{A}$ and A have the same K-theory).

The basic commutative example of a p-summable cycle is the following:

- $A = C(M)$ is the algebra of continuous functions on a smooth compact manifold M,
- H is the Hilbert space of L^2 sections of a bundle over M, and F is an elliptic classical pseudodifferential operator of order 0.

In this case, for a smooth function $f \in C(M)$, the commutator $[f, F]$ is a pseudodifferential operator of order 0 with a zero principal symbol. Hence it has order -1. As any pseudodifferential operator with order $< -n$ (where n is the dimension of M) is trace class, we deduce that $[f, F] \in \mathcal{L}^p(H)$ for any $p > n$. Hence the cycle (H, F) is p-summable.

It is not true that any K-homology class can be represented by a p-summable cycle. The entire cyclic cohomology of [12] allows us to give an index formula in the form of a convergent series for more general cocycles (see also [29]).

2 The Local Index Formula (cf. [19])

It may be useful to give an index formula other than the formula (1) above. In the example of a pseudodifferential operator the latter has several unpleasant aspects:

1. The natural operators in index theory problems (e.g. signature operator, Dirac operator, $\bar{\partial}$-operator,...) are differential operators of order 1. We can change them into order 0 operators, but this add technical difficulties, since we then have to deal with pseudodifferential operators.
2. This formula involves the operator trace: the cyclic cocycle is thus sensitive to a modification of F by a smoothing operator, unlike its cyclic cohomology class. In other words, the formula is not local.

Because of this, Connes and Moscovici gave a formula for the cocycle which is local, though more intricate at first glance; it fits into an abstract setting modelled on (pseudo)differential operators on a manifold.

Definition 2.1. *([14]) A spectral triple is a triple $(H, \mathcal{A}, D)$ where*

- *H is a $\mathbb{Z}/2\mathbb{Z}$-graded Hilbert space,*
- *$\mathcal{A}$ is an involutive subalgebra of the algebra $\mathcal{L}(H)^{(0)}$ of bounded linear operators on H preserving the grading,*
- *D is an injective (unbounded) selfadjoint operator on H of degree 1 with respect to the grading, such that:*
  1. *the algebra $\mathcal{A}$ is contained in the domain of the unbounded derivation $x \mapsto [D, x]$ (defined on a subalgebra of $\mathcal{L}(H)$), and also in the domains of all the iterated composites of the unbounded derivation $x \mapsto [|D|, x]$;*
  2. *the operator $|D|^{-1}$ lies in a Schatten class;*
  3. *if P is in the smallest subalgebra of $\mathcal{L}(H)$ containing $\mathcal{A}$, the elements of the form $[D, a]$, $a \in \mathcal{A}$, the grading operator γ, and which is stable under commutation with $|D|$, then the function $z \mapsto \mathrm{Tr}(P|D|^{-z})$ has an analytic continuation on the complement of a subset S of $\mathbb{C}$ called dimension spectrum (which is discrete in general).*

We say that the dimension spectrum is simple if there are only simple pole singularities in S.

Condition 1 above means that:

- the domain of D is invariant under the action of $\mathcal{A}$, and for any $a \in \mathcal{A}$ the operator $[D, a]$ is bounded, i.e. extends to a continuous operator on H;
- for any $a \in \mathcal{A}$ the map $t \mapsto e^{it|D|} a e^{-it|D|}$ from $\mathbb{R}$ to $\mathcal{L}(H)$ is smooth (with respect to the norm topology).

If $\mathcal{A}$ is not unital, we replace condition 2 by:

2'. there exists $p \in \mathbb{R}_+^*$ such that, for any $a \in \mathcal{A}$, the operator $a|D|^{-p}$ is trace class.

The above conditions are precisely modelled on those satisfied by an elliptic (pseudo)differential operator of order 1 on a compact manifold. They are also satisfied in other contexts such as Fuchs type operators on a manifold with conical singularities (cf. [8, 35, 36]).

Let $(H, \mathcal{A}, D)$ be a spectral triple. We can define the Sobolev spaces for D as follows. For $s \geq 0$, H^s is the domain of $|D|^s$; for $s \leq 0$, H^s is the completion of H for the norm $x \mapsto \|x\|_s = \|(|D|^s x)\|$. Further, H^∞ is the smooth domain of D, that is the intersection of all the Sobolev spaces and $H^{-\infty}$ is the union of the spaces H^s. Note by condition 1 each element of $\mathcal{A}$ preserves each space H^s. Finally, denote by $\mathcal{P}$ the algebra of pseudodifferential operators, that is the algebra generated by $\mathcal{A}$, D and $|D|^{-1}$; it acts naturally on $H^{-\infty}$ and leaves H^∞ invariant.

For $a \in \mathcal{P}$, let $\delta(a) = [D^2, a]$ and set $a^{(i)} = \delta^i(a)$, $i \in \mathbb{N}$.

For $P \in \mathcal{P}$, let us denote by $\fint P$ the residue at 0 of the meromorphic function which, for $\operatorname{Re} s$ large enough, equals $\operatorname{Tr}(\gamma P |D|^{-s})$.

Theorem 2.2. *Let $(H, \mathcal{A}, D)$ be a spectral triple. Assume that the dimension spectrum is simple and discrete.*

a) For $a \in \mathcal{A}$, let $\varphi_0(a)$ be the residue at 0 of the function $z \mapsto z^{-1} \operatorname{Tr}(\gamma a |D|^{-z})$. For $n \in \mathbb{N}$ non-zero set

$$\varphi_n(a_0, a_1, \ldots, a_n) = \sum_{k \in \mathbb{N}^n} c_{n,k} \fint a_0 [D, a_1]^{(k_1)} \ldots [D, a_n]^{(k_n)} |D|^{-n-2|k|}$$

where $|k| = \sum_j k_j$ and

$$c_{n,k} = \frac{(-1)^{|k|} \, 2 \, \Gamma\!\left(|k| + \frac{n}{2}\right)}{k! \, (k_1 + 1)(k_1 + k_2 + 2) \ldots (k_1 + \ldots + k_n + n)}.$$

The family $\varphi = (\varphi_{2p})_{p \in \mathbb{N}}$ is a cocycle in the (b, B)-bicomplex.

Let $F = D|D|^{-1}$ be the sign of D.

b) The class of φ in the (b, B)-bicomplex coincides with the class of the cyclic cocycle $(a_0, a_1, \ldots, a_n) \mapsto \operatorname{Tr}(\gamma a_0 [F, a_1] \ldots [F, a_n])$.

Let us make few comments on the above formula:

- For any $n \in \mathbb{N}$ and any $k \in \mathbb{N}^n$, the operator
$$a_0[D, a_1]^{(k_1)} \ldots [D, a_n]^{(k_n)}|D|^{-|k|}$$
is bounded. Therefore, for $n + |k| \geq p$, the operator
$$a_0[D, a_1]^{(k_1)} \ldots [D, a_n]^{(k_n)}|D|^{-n-2|k|}$$
is trace class, so the residue $\displaystyle\int a_0[D, a_1]^{(k_1)} \ldots [D, a_n]^{(k_n)}|D|^{-n-2|k|}$ is zero; the index formula has thus a finite number of terms.

- We have assumed that D is invertible. In practice, it may happen that the input operator has a finite dimensional kernel. In this case, we interpret $|D|^{-s}$ as being zero on the kernel of D. Using this convention all the terms in the formula are left unchanged, except for φ_0 to which we must add a term $a \mapsto \mathrm{Tr}(\gamma a p)$ where p is the orthogonal projection onto $\ker D$.

- In all the known natural examples the dimension spectrum is simple. We can generalise this formula to non-simple dimension spectra: we only need to reinterpret the terms by replacing
$$\Gamma\!\left(|k| + \frac{n}{2}\right) \int a_0[D, a_1]^{(k_1)} \ldots [D, a_n]^{(k_n)}|D|^{-n-2|k|}$$
by the residue at $s = |k| + \frac{n}{2}$ of the function
$$z \mapsto \Gamma(z)\,\mathrm{Tr}\!\left(\gamma a_0[D, a_1]^{(k_1)} \ldots [D, a_n]^{(k_n)}|D|^{-2z}\right);$$
this yields a sum of terms in which there appear derivatives of the Γ-function, and hence values of the ζ-function. We might have hoped to use these formulas and the integrability of the index in order to get algebraic equations involving $\zeta(2k+1)$, but instead we use a renormalisation procedure to replace these terms by rational ones without having any effect on the cohomology class (cf. [19]).

3 The Diff-Invariant Signature Operator (cf. [19])

Let (V, F) be a foliated manifold. The transverse bundle does not always have a holonomy invariant Hermitian structure, thus there does not exist in general a transversally elliptic pseudodifferential operator of order $\neq 0$ with a holonomy invariant principal symbol. Having picked a transversal the problem is to construct an elliptic operator invariant under a pseudogroup Γ of diffeomorphisms of M. Such an operator does not exist in general: the group of diffeomorphisms fixing the principal symbol must fix a metric.

This problem can be solved by using the bundle of metrics on the tangent bundle of M. This idea originally comes from [11], where it was used in order to construct cyclic cocycles having a continuation on the foliation C^*-algebra, and was also used in [28] to build almost invariant order 0 pseudodifferential operators.

Let M be an oriented compact manifold of dimension n. Denote by R the total space of the positive frame bundle of M. The group $GL^+(n)$ acts naturally on R and we have $M = R/GL^+(n)$. Denote by $P = R/SO(n)$ the space of metrics on M.

Let Γ be a group of orientation preserving diffeomorphisms of M; it acts naturally on R and on P. This action preserves the fibration $p: P \to M$; the tangent bundle TP of P thus contains a vertical Γ-invariant subbundle L. Moreover, the bundle L and the quotient TP/L both have Γ-invariant metrics: the metric on L comes from a $GL^+(n)$-invariant metric on $GL^+(n)/SO(n)$; the bundle TP/L is isomorphic to $p^*(TM)$ and a point of P *is* a metric on this bundle.

3.1 The Hilbert space $\mathcal{H}$ and the algebra $\mathcal{A}$

In the construction of the signature operator on P we will make a distinction between the vertical direction (along the fibers of the fibration $P \to M$) and the horizontal direction (transverse to the fibration). This operator will not act on the bundle $\Lambda(T^*P)$ of forms on P, but rather on the tensor product $p^*(\Lambda(T^*M)) \otimes \Lambda L^*$ of the space of forms on M with the exterior algebra of the dual bundle of L. Note this is a Hermitian bundle, unlike the bundle $\Lambda(T^*P)$.

The bundles $p^*(\Lambda(T^*M))$ and ΛL^* have grading operators γ_H and γ_V. As for the classical signature operator these are given by a $*$-operator defined up to a phase factor by the equality $\langle *\omega, \omega' \rangle v = \omega \wedge \omega'$ for ω, ω' real forms, where v is the positive volume form with norm 1 (using the orientations of the bundles).

The Hilbert space $\mathcal{H}$ is the space of L^2 sections of this bundle with respect to the Γ-invariant measure on P given by the volume form $v_V \otimes v_H$. The grading operator on our bundle, and so on $\mathcal{H}$, is $\gamma = \gamma_V \otimes \gamma_H$.

The algebra $\mathcal{A}$ is the crossed product $C_c^\infty(P) \rtimes \Gamma$. It is generated by operators of the form fU_g, where:

- $f \in C_c^\infty(P)$ is a compactly supported function on the total space P seen as a multiplication operator on $\mathcal{H}$;
- $g \in \Gamma$ and $g \mapsto U_g$ is the natural unitary representation of Γ on $\mathcal{H}$ given by the formula

$$(U_g \omega)(x) = g \cdot \omega(g^{-1}x), \qquad x \in P.$$

The main algebraic formula satisfied by these operators is $U_g f = (g \cdot f) U_g$, where $(g \cdot f)(x) = f(g^{-1}(x))$.

3.2 The vertical part of the signature operator

At a point x of M we consider the signature operator $d_V + d_V^*$ on the manifold $P_x = p^{-1}(x)$, with coefficients in the trivial bundle $E = \Lambda(T_x^* M)$. This yields a Γ-invariant differential operator of order 1.

Note that the bundle E is trivial but its Hermitian structure is not constant. So the operator $d_V^* + \gamma d_V \gamma$ is not zero, but has order 0. In this sense, the operator $d_V + d_V^*$ anticommutes with γ and thus has degree 1 with respect to the grading.

For a reason which will be made clearer below, we also need to consider a vertical signature operator of order 2: we take $Q_V = d_V^* d_V - d_V d_V^*$. This anticommutes with γ and its principal symbol defines the same K-theory class as the one of $d_V + d_V^*$.

3.3 The horizontal part of the signature operator

This operator is $Q_H = d_H + d_H^*$. Nevertheless, to build it, we need to use a connection, that is a section $s\colon TP/L \to TP$. It yields an isomorphism $T^*P \simeq L^* \oplus p^*(T^*M)$, hence an isomorphism $\Lambda(T^*P) \simeq p^*(\Lambda(T^*M)) \otimes \Lambda L^*$ that we will use to construct d_H. Note that neither the section s nor d_H are Γ-invariant. On the other hand, the horizontal symbol of d_H, that is the restriction of its principal symbol to $p^*(\Lambda(T^*M)) \subset T^*P$, is invariant. Indeed, the principal symbol d_H at a point $(y, \xi) \in T^*P$ ($y \in P$, $\xi \in T_y^* P$) is

$$\omega_1 \otimes \omega_2 \mapsto ({}^t(s \circ p)(\xi) \wedge \omega_1) \otimes \omega_2$$

for $\omega_1 \in \Lambda(T_{p(y)}^* M)$ and $\omega_2 \in \Lambda(L_y^*)$.

3.4 The spectral triple

Let $Q = Q_V + Q_H = (d_V^* d_V - d_V d_V^*) + \gamma_V (d_H + d_H^*)$. One can show that Q is self-adjoint (this is not so easy since Q is not elliptic and P is not compact).

For $f \in C_c^\infty(P)$, the operator $[f, Q_H]$ is differential of order 0: it is thus bounded. The operator $[f, Q_V]$ is differential of order 1, but its horizontal symbol is zero. Hence it is a longitudinal operator along the fibration $P \to M$.

The operator Q_V is Γ-invariant, and so for any $g \in \Gamma$ we have $[U_g, Q_V] = 0$. The operator $[U_g, Q_H]U_g^{-1}$ is differential of order 1 and its horizontal symbol is zero. As Q_V is longitudinally elliptic of order 2, the longitudinal differential operators of order 1 (with compact support) are dominated by $|Q_V|^{1/2}$, and thus by $|Q|^{1/2}$.

By functional calculus we define an operator D such that $Q = D|D|$. By the previous discussions, and using the equality

$$D = \frac{\sqrt{2}}{2\pi} \int_0^{+\infty} Q(Q^2 + \mu)^{-1} \mu^{-1/4} d\mu,$$

one can check that, for any $h \in \mathcal{A}$, the commutator $[h, D]$ is bounded.

Theorem 3.1. *[19] The triple $(\mathcal{A}, \mathcal{H}, D)$ is a spectral triple.*

The proof this theorem relies heavily on the hypoelliptic calculus on Heisenberg manifolds (the latter are manifolds endowed with a subbundle of the tangent bundle, cf. [5]). The interesting subbundle here is the (integrable) subbundle tangent to the fibers of the fibration p. Note that we meet several difficulties here due to the non-compactness of P.

Finally, it can be proved (by means of [5]) that the natural hypoelliptic operators on a not necessarily integrable Heisenberg manifold also give rise to spectral triples (cf. [38]).

4 The "Transverse" Hopf Algebra

Let M be a manifold and Γ a group of diffeomorphisms of M. We are provided with a signature operator D acting on the total space of the bundle P of metrics on M. We thus deduce a local index formula. The problem is that the actual computation of this formula is a rather formidable task: even in the case where M is of dimension 1, so that the space P is of dimension 2 and D is 3-summable (i.e. $f|D|^{-z}$ is of trace class for $f \in C_c^\infty(P)$ if $\operatorname{Re} z > 3$), this formula gives thousands of terms to compute. If each one of these is relatively easy to deal with, it is absolutely impossible to write a satisfactory proof using these formulas when $n \geq 2$!

For that, Connes and Moscovici construct a Hopf algebra $\mathcal{H}_n$, depending only on the dimension n of M. This plays the role of a group of symmetries and makes it possible to eliminate several unnecessary terms from the beginning. Moreover, as the transverse signature operator comes from the action of $\mathcal{H}_n$, the cyclic cocycle given by the local index formula associated to this operator comes from a cocycle in the cyclic cohomology of the Hopf algebra $\mathcal{H}_n$, using a "classifying" map

$$\operatorname{HC}^*_{\operatorname{Hopf}}(\mathcal{H}_n) \to \operatorname{HC}^*(\mathcal{A}),$$

associated to the action of $\mathcal{H}_n$ on $\mathcal{A}$.

The cyclic cohomology of $\mathcal{H}_n$ is calculated using a van Est type theorem: it is the Gel'fand-Fuchs cohomology $H^*(WO(n))$.

A Hopf algebra of exactly the same type as the algebra $\mathcal{H}_1$ of Connes-Moscovici was constructed by Kreimer ([33]), then improved by Connes-Kreimer ([16–18]).

The construction of this Hopf algebra is similar to Kac's construction of the Hopf algebra associated to a matched pair of groups, which we shall now describe.

4.1 Matched pairs of groups (cf. [30])

Let G be a (finite) group and G_1, G_2 two subgroups such that the function $(x, y) \mapsto xy$ is a bijection from $G_1 \times G_2$ onto G. It is then possible to build a Hopf algebra in the following way.

- By identifying G_2 with $G_1\backslash G$ and G_1 with G/G_2, there are natural actions of G on G_2 (on the right) and on G_1 (on the left).
- As an algebra, $\mathcal{H}$ is isomorphic to the crossed product $C(G_1) \rtimes G_2$ (for the pointwise product) of the functions on G_1 by the action of G_2.
- The algebra which is dual to coalgebra structure of $\mathcal{H}$ is the crossed product $\widehat{\mathcal{H}} = C(G_2) \rtimes G_1$.

The duality between these is written

$$\langle \varphi U_s | U_g f \rangle = f(s)\varphi(g),$$

for $s \in G_1$, $g \in G_2$, $f \in C(G_1)$ and $\varphi \in C(G_2)$.

- The coproduct $\Delta : \mathcal{H} \to \mathcal{H} \otimes \mathcal{H}$ for $\mathcal{H}$ is written as follows: for $f \in C(G_1)$, $\Delta(f) \in C(G_1 \times G_1) = C(G_1) \otimes C(G_1) \subset \mathcal{H} \otimes \mathcal{H}$ is the function $(s,t) \mapsto f(st)$; for $g \in G_2$ the formula for $\Delta(U_g)$ is slightly more complicated: it is written

$$\Delta(U_g) = \sum_{s \in G_1} U_g \delta_s \otimes U_{p(gs)},$$

where $\delta_s \in C(G_1)$ is the characteristic function of s, and for $x \in G$ we denote by $p(x) \in G_2$ the unique element such that $xp(x)^{-1} \in G_1$.
- The antipode $S : \mathcal{H} \to \mathcal{H}$ is the antihomomorphism given by $S(\delta_s U_g) = U_{p(gs)} \delta_{s^{-1}}$ where $s \in G_1$ and $g \in G_2$.

Remarks.

1. The above example of a "quantum group" can also be twisted by a cocycle.
2. This example has been studied by several authors and has been generalized to the case where G is a locally compact group and G_1, G_2 are two closed subgroups such that the map $(s,g) \mapsto sg$ is a homeomorphism from $G_1 \times G_2$ onto G, or even onto a dense subset of G (cf. [2, 3, 37, 41, 43]).

Note that the formula for the coproduct of $\mathcal{H}$ extends to the algebra $\mathcal{R} = C(G_1) \rtimes G \supset \mathcal{H}$: define $\Delta_\mathcal{R} : \mathcal{R} \to \mathcal{R} \otimes \mathcal{H}$ by $\Delta_\mathcal{R}(f) = \Delta(f) \in \mathcal{H} \otimes \mathcal{H} \subset \mathcal{R} \otimes \mathcal{H}$ for $f \in C(G_1)$, and for $g \in G$ let

$$\Delta_\mathcal{R}(U_g) = \sum_{s \in G_1} U_g \delta_s \otimes U_{p(gs)}.$$

The extended function is again a homomorphism of algebras, and again satisfies the associativity condition $(\Delta_\mathcal{R} \otimes \mathrm{id}) \circ \Delta_\mathcal{R} = (\mathrm{id} \otimes \Delta) \circ \Delta_\mathcal{R}$, hence it is a coaction of the Hopf algebra $\mathcal{H}$ on the algebra $\mathcal{R}$. Note that here $\mathcal{R}$ can be replaced by the algebra $C(G_1) \rtimes \Gamma$ for any subgroup Γ of G.

An equivalent way to view the coaction of $\mathcal{H}$ on $\mathcal{R}$ is through the dual Hopf algebra $\widehat{\mathcal{H}} = C(G_2) \rtimes G_1$: for $\alpha \in \widehat{\mathcal{H}}$ and $x \in \mathcal{R}$ we set $\alpha \cdot x = (\mathrm{id} \otimes \alpha)\Delta_\mathcal{R}(x) (\in \mathcal{R})$. Since the product of $\widehat{\mathcal{H}}$ is dual to the coproduct of $\mathcal{H}$, the co-associativity of $\Delta_\mathcal{R}$ implies that it is an action, i.e. $(\alpha\beta) \cdot x = \alpha \cdot (\beta \cdot x)$ for $\alpha, \beta \in \widehat{\mathcal{H}}$ and

$x \in \mathcal{R}$. As the coproduct $\hat{\Delta}$ on $\hat{\mathcal{H}}$ is dual to the product on $\mathcal{H}$, the fact that $\Delta_{\mathcal{R}}$ is a homomorphism amounts to the formula

$$m\big(\hat{\Delta}(\alpha) \cdot (x \otimes y)\big) = \alpha \cdot (xy) \qquad (2)$$

(for $\alpha \in \hat{\mathcal{H}}$ and $x, y \in \mathcal{R}$, and with $m : \mathcal{R} \otimes \mathcal{R} \to \mathcal{R}$ denoting the multiplication for $\mathcal{R}$).

On the generators of $\hat{\mathcal{H}}$ we obtain the following formula:

$$(U_s \varphi) \cdot (U_g f) = U_g \varphi\big(p(gs)\big) f^s \qquad (3)$$

for $s \in G_1$, $g \in G$, $f \in C(G_1)$ and $\varphi \in C(G_2)$, where f^s denotes the function $t \mapsto f(ts)$.

4.2 The Hopf algebra $\mathcal{H}_n$ (cf. [20])

The example of matched pairs will serve as a template for constructing the Hopf algebra $\mathcal{H}_n$. Here the group G is the group of all orientation preserving diffeomorphisms of $\mathbb{R}^n$; the subgroup G_1 will be the group $\mathbb{R}^n \rtimes GL^+(n)$ of affine diffeomorphisms; the subgroup G_2 will be the diffeomorphisms of $\mathbb{R}^n$ such that $f(0) = 0$ and $df_0 = \mathrm{id}$. Note that G/G_2 is the space of frames on $\mathbb{R}^n$.

The Hopf algebra $\mathcal{H}_n$ can be considered as a kind of crossed product of the algebra of functions on G_2, which are polynomials in the finite order jets of a diffeomorphism, by the enveloping algebra of the Lie algebra of G_1. This can actually be seen from its action on the algebra $\mathcal{R} = C_c^\infty(G_1) \rtimes G$, using formulas (2) and (3) to take into account the action and to define the coproduct on $\mathcal{H}_n$. Note that the Hopf algebra was originally built "by hand" using its action on $\mathcal{R}$. The link with the matched pairs of groups clarifies several calculations.

The action of $GL^+(n)$ commutes with that of G: it follows that if Y is an element of the Lie algebra of $GL^+(n)$, then $\Delta(Y) = Y \otimes 1 + 1 \otimes Y$. Let X_i be a generator for the lie algebra of $\mathbb{R}^n$. Letting X_i act on a product fU_g, we find that $X_i(fU_g) = X_i(U_g(f^g)) = U_g(X_i(f^g)) = (X_i(f^g))^{g^{-1}} U_g$. Comparing with $X_i(f)U_g$ we see that the coproduct of X_i is of the form $\Delta(X_i) = X_i \otimes 1 + 1 \otimes X_i + \sum Y_{j,k} \otimes \delta_{i,j,k}$, where $\delta_{i,j,k}$ are functions on G_2. The coproduct of these elements is given by $\Delta(\delta_{i,j,k}) = \delta_{i,j,k} \otimes 1 + 1 \otimes \delta_{i,j,k}$.

The elements $X_i, Y_{i,j}$ and $\delta_{i,j,k}$ generate the Hopf algebra $\mathcal{H}_n$. In particular they can be used to construct new elements

$$\delta_{I,\alpha} = \Big[X_{i_1}, \big[X_{i_2}, \ldots [X_{i_k}, \delta_\alpha] \ldots \big]\Big],$$

$I = (k, i_1, \ldots, i_k)$, $\alpha \in \{1, \ldots, n\}^3$. These elements act by multiplication on $\mathcal{R}$ and they form the commutative algebra of functions on G_2.

The algebra $\mathcal{H}_n$ is the vector space generated by the products xy where x is an element of the enveloping algebra of the Lie algebra of G_1 and y is an element of the algebra of polynomials in the variables $\delta_{I,\alpha}$. The coproduct on $\mathcal{H}_n$ is defined inductively on the generators. It has already been observed that $\Delta(\delta_\alpha) = \delta_\alpha \otimes 1 + 1 \otimes \delta_\alpha$ for $\alpha \in \{1,\ldots,n\}^3$. The co-unit ε is the unique character which is zero on the X_i, $Y_{i,j}$ and δ_α. The antipode is the unique anti-automorphism S of $\mathcal{H}_n$ such that $S(Y_{i,j}) = -Y_{i,j}, S(\delta_\alpha) = -\delta_\alpha$ and $S(X_i) = -X_i + \sum_{j,k} \delta_{i,j,k} Y_{j,k}$

4.3 Cyclic cohomology of Hopf algebras (cf. [21, 22])

Let $\mathcal{H}$ be a Hopf algebra (in the sense of [40]). Let Δ denote the coproduct, $\varepsilon\colon \mathcal{H} \to \mathbb{C}$ the co-unit and $S\colon \mathcal{H} \to \mathcal{H}$ the antipode. In general S^2 need not be the identity. A *modular pair* $(F, \widehat{F})$ must therefore be given, where $F \in \mathcal{H}$ is a group-like element, that is $\Delta(F) = F \otimes F$, and $\widehat{F}\colon \mathcal{H} \to \mathbb{C}$ is a character, i.e. a homomorphism of algebras. Define $\widetilde{S} = (\widehat{F} \otimes S) \circ \Delta\colon \mathcal{H} \to \mathcal{H}$. Suppose that $\widehat{F}(F) = 1$ and

$$\widetilde{S}^2(h) = FhF^{-1} \qquad \text{for any } h \in \mathcal{H}. \tag{4}$$

Then we have the following formulas:

- $\Delta \circ \widetilde{S} = (S \otimes \widetilde{S}) \circ \sigma \circ \Delta$, where $\sigma : \mathcal{H} \otimes \mathcal{H} \to \mathcal{H} \otimes \mathcal{H}$ is the function $x \otimes y \mapsto y \otimes x$.
- $\varepsilon \circ \widetilde{S} = \widehat{F}$ and $\widehat{F} \circ \widetilde{S} = \varepsilon$.

Remark. One such modular pair is naturally constructed in the framework of the locally compact quantum groups of [34]: in this context F and $\widehat{F}$ represent the Radon-Nykodym derivatives of the right Haar measure with respect to the left Haar measure, for respectively $\mathcal{H}$ and its dual $\widehat{\mathcal{H}}$. In this case however $\widehat{F}(F) \neq 1$. Some minor modifications are then required as follows. It is in fact always possible to reduce to the case where $\widehat{F}(F) = 1$ using a "modular square" construction. In the case of matched pairs of groups, however, the modular pairs always satisfy the property $\widehat{F}(F) = 1$ (cf. [3, 43]).

Let $\mathcal{H}$ be a Hopf algebra provided with a modular pair $(F, \widehat{F})$. There are cyclic cohomology groups naturally associated to it, given by a bicomplex $(C_{p,q}, b, B)$ defined as follows.

- For $p \geq q \geq 0$, let $C_{p,q} = \mathcal{H}^{\otimes p-q}$.
- For $k \in \mathbb{N}$ and $0 \leq j \leq k+1$, define functions $b_j : \mathcal{H}^{\otimes k} \to \mathcal{H}^{\otimes k+1}$ by setting $b_0(h) = 1 \otimes h$, $b_{k+1}(h) = h \otimes F$ and, for $1 \leq j \leq k$, $b_j = \mathrm{id}_{\mathcal{H}^{\otimes j-1}} \otimes \Delta \otimes \mathrm{id}_{\mathcal{H}^{\otimes k-j}}$. Finally let $b = \sum_{j=0}^{k+1}(-1)^j b_j$.
- We will also make use of the functions $\Delta^j : \mathcal{H} \to \mathcal{H}^{\otimes j+1}$ defined by $\Delta^1 = \Delta$, $\Delta^2 = (\Delta \otimes \mathrm{id}_\mathcal{H}) \circ \Delta = (\mathrm{id}_\mathcal{H} \circ \Delta) \circ \Delta$ (by coassociativity) and $\Delta^{j+1} = (\Delta^j \otimes \mathrm{id}_\mathcal{H}) \circ \Delta = (\mathrm{id}_\mathcal{H} \circ \Delta^j) \circ \Delta$.

- The cyclic permutation $\lambda: \mathcal{H}^{\otimes k} \to \mathcal{H}^{\otimes k}$ is defined by

$$\lambda(h_1 \otimes \ldots \otimes h_k) = (-1)^k \big(\Delta^{k-1} \circ \widetilde{S}(h_1)\big)\big(h_2 \otimes \ldots \otimes h_k \otimes F\big).$$

From (4) it follows that λ^{k+1} is the identity.
- The function $B: \mathcal{H}^{\otimes k+1} \to \mathcal{H}^{\otimes k}$ is the composition $A \circ u \circ D$ where

$$D = \mathrm{id} - \lambda: \mathcal{H}^{\otimes k+1} \to \mathcal{H}^{\otimes k+1}, \quad A = \sum_{j=0}^{k} \lambda^j : \mathcal{H}^{\otimes k} \to \mathcal{H}^{\otimes k}$$

and $u(h_1 \otimes \ldots \otimes h_{k+1}) = \big(\Delta^{k-1} \circ \widetilde{S}(h_1)\big)\big(h_2 \otimes \ldots \otimes h_{k+1}\big)$.

Definition 4.1. *(cf. [21, 22] - cf. also [24]) The cyclic cohomology of a Hopf algebra $\mathcal{H}$ is the cohomology $\mathrm{HC}^*_{\mathrm{Hopf}}(\mathcal{H})$ of the bicomplex (b, B) defined above.*

4.4 The classifying map

The maps b and B in the above bicomplex are built precisely to allow the construction of a classifying map. Thus let $\mathcal{R}$ be an algebra equipped with an action of $\mathcal{H}$. Suppose an F-*trace* on $\mathcal{R}$ is given, that is a map $\tau: \mathcal{R} \to \mathbb{C}$ satisfying $\tau(xy) = \tau(y(F \cdot x))$ for $x, y \in \mathcal{R}$. We will say that τ is $\widehat{F}$-*invariant* if $\tau(h \cdot x) = \widehat{F}(h)\tau(x)$ for $x \in \mathcal{R}$ and $h \in \mathcal{H}$, or equivalently $\tau\big((h \cdot x)y\big) = \tau\big(x(\widetilde{S}(h) \cdot y)\big)$ for $x, y \in \mathcal{R}$ and $h \in \mathcal{H}$.

Let τ be an $\widehat{F}$-invariant F-trace. The map which associates to $H = h_1 \otimes \ldots \otimes h_k \in \mathcal{H}^{\otimes k}$ the form $\varphi_H: (a_0, \ldots, a_k) \mapsto \tau\big(a_0(h_1 \cdot a_1) \ldots (h_k \cdot a_k)\big)$ is, by construction, a morphism of bicomplexes. Hence it gives rise to a homomorphism,

$$\mathrm{HC}^*_{\mathrm{Hopf}}(\mathcal{H}) \to \mathrm{HC}^*(\mathcal{R}),$$

called the *classifying map*.

4.5 Cyclic cohomology of the Hopf algebra $\mathcal{H}_n$ (cf. [20])

In the case of a bicrossed product (of locally compact groups), the modular pair $(F, \widehat{F})$ is written in terms of the modular functions δ_1, δ_2 and δ for the groups G_1, G_2 and G (cf. [3, 43]). The elements $F, \widehat{F}$ belong respectively to $C(G_1) \subset \mathcal{H}$ and $C(G_2) \subset \widehat{\mathcal{H}}$ and are given by $F(s) = \delta_1(s)\delta(s)^{-1/2}$ and $\widehat{F}(s) = \delta_2(s)\delta(s)^{-1/2}$.

For the Hopf algebra $\mathcal{H}_n$ (which is the algebra $\widehat{\mathcal{H}}$ of the matched pair G_1, G_2 described above), the group G_2 is "nilpotent" and the group G is simple, hence the functions δ_2 and δ are identically equal to 1. It follows that $F = 1$ and $\widehat{F}$ is the modular function of G_2. Thus $\widehat{F}(X_i) = \widehat{F}(\delta_\alpha) = 0$ and $\widehat{F}(Y_{i,j}) = \mathrm{Tr}(Y_{i,j}) = \delta_i^j$.

A van Est type theorem can be used to calculate the periodic cyclic cohomology $\mathrm{HC}^*_{\mathrm{per,Hopf}}(\mathcal{H}_n)$ of the Hopf algebra $\mathcal{H}_n$: it is the Gel'fand-Fuchs cohomology of the infinite dimensional Lie algebra $\mathfrak{g}$ of G.

Theorem 4.2. *(cf. [20]) For $j = 0, 1$ there is a morphism of complexes which induces an isomorphism*

$$\bigoplus_{i \in \mathbb{N}} H^{j+2i}(\mathfrak{g}) \to \mathrm{HC}^j_{\mathrm{per,Hopf}}(\mathcal{H}_n).$$

In the case of the algebra $\mathcal{H}_n$ acting on $\mathcal{R} = C_c^\infty(G_1) \rtimes G$, the linear functional defined by $f \mapsto \int_{G_1} f(s)\,ds$ and $fU_g \mapsto 0$ if $g \neq \mathrm{id}$ is an $\widehat{F}$-invariant trace. This gives rise to homomorphisms

$$\mathrm{HC}^*_{\mathrm{Hopf}}(\mathcal{H}_n) \to \mathrm{HC}^*(\mathcal{R}) \quad \text{and} \quad \bigoplus_{i \in \mathbb{N}} H^{j+2i}(\mathfrak{g}) \simeq \mathrm{HC}^j_{\mathrm{per,Hopf}}(\mathcal{H}_n) \to \mathrm{HC}^j_{\mathrm{per}}(\mathcal{R}).$$

However, the interesting algebra is not $\mathcal{R}$, but rather $\mathcal{A}_n = C_c^\infty(P_n) \rtimes G$ where $P_n = G_1/SO(n)$ is the space of metrics. The algebra $\mathcal{A}_n$ is the algebra of fixed points of $\mathcal{R}$ under the action of the group $SO(n)$. We use here a relative classifying map

$$\bigoplus_{i \in \mathbb{N}} H^{j+2i}(\mathfrak{g}; SO(n)) \simeq \mathrm{HC}^j_{\mathrm{per,Hopf}}(\mathcal{H}_n; SO(n)) \to \mathrm{HC}^j_{\mathrm{per}}(\mathcal{A}_n).$$

The cohomology group $H^k(\mathfrak{g}; SO(n))$ is in fact the Gel'fand Fuchs cohomology $H^*(W\,SO(n))$ (cf. [25,26]). It is generated by Pontrjagin classes and secondary classes.

Let us finally return to the case in which we were originally interested: Let M be a manifold and Γ a pseudogroup of orientation preserving diffeomorphisms of M. Let $P = P_M$ be the total space of the bundle of metrics on M. The manifold M can be replaced by an open cover $(U_i)_{i \in I}$ of M and the pseudogroup Γ by a pseudogroup Γ' of local diffeomorphisms of $\Omega = \coprod_{i \in I} U_i$. The algebra $C_c^\infty(P_\Omega) \rtimes \Gamma'$ obtained from this is Morita equivalent to $C_c^\infty(P_M) \rtimes \Gamma$, so that the two algebras have the same cyclic cohomology. Since Ω is (diffeomorphic to) an open subset of $\mathbb{R}^n$, we get the desired classifying homomorphism:

$$\bigoplus_{i \in \mathbb{N}} H^{j+2i}(\mathfrak{g}; SO(n)) \to \mathrm{HC}^j_{\mathrm{per}}(C_c^\infty(P_M) \rtimes \Gamma).$$

4.6 The cyclic cocycle Φ (cf. [20])

Let ϕ be the cyclic cocycle from the local index formula for the transverse signature operator D. We must now show that ϕ is the image under the classifying map of some cyclic cocycle Φ in the cohomology $\mathrm{HC}^*_{\mathrm{Hopf}}(\mathcal{H}_n, SO(n))$. For this we use the following observations.

1. The operator D (which is an order 1 pseudodifferential operator in the Heisenberg calculus) can be replaced in the index formula by the second order differential operator $Q = D|D|$; the cyclic cohomology class is not affected by this.

2. The operator Q comes from the enveloping algebra $\mathcal{U}(\mathfrak{g}_1)$ of the lie algebra of $G_1 = \mathbb{R}^n \rtimes GL^+(n)$, which is contained in $\mathcal{H}_n$. It is thanks to this observation that Q can be easily shown to be self-adjoint. The cocycle ϕ is thus written as a combination of terms

$$(a_0, \ldots, a_k) \mapsto \int a_0 R_1 a_1 \ldots R_k a_k T,$$

where $R_j \in \mathcal{U}(\mathfrak{g}_1)$ and T is a pseudodifferential operator in the Heisenberg calculus.

3. By successively commuting the vector fields in the R_j with the elements of $\mathcal{A}$, we can reduce to terms of the form

$$(a_0, \ldots, a_k) \mapsto \int a_0 (h_1 \cdot a_1) \ldots (h_k \cdot a_k) T,$$

where T is a pseudodifferential operator in the Heisenberg calculus.

4. Suppose that the (pseudo-)group Γ acts freely on the frames of $\mathbb{R}^n$. Then all terms of the form $\int (f_k U_{g_0})(h_1 \cdot (f_1 U_{g_1})) \ldots (h_k \cdot (f_k U_{g_k})) T$ where $g_0 g_1 \ldots g_k \neq 1$ will cancel.

5. If $f \in C_c^\infty(P)$, then the residue $\int fT$ may be written $\int_P f\alpha$ where α is a volume form on P obtained by integrating the order $-n(n+3)/2$ symbol of T (in the Heisenberg calculus $n(n+3)/2$ is the critical dimension) over the "sphere". As the operator Q is invariant under the action of G_1, the operator T from step 3 is also invariant under the action of G_1, thus α is the volume form. This shows that ϕ can be expressed as a sum of terms

$$(a_0, \ldots, a_k) \mapsto \tau(a_0, \ldots, a_k) \mapsto \tau\big(a_0(h_1 \cdot a_1) \ldots (h_k \cdot a_k)\big).$$

6. The above results show that ϕ is the image of an element Φ under the classifying homomorphism of bicomplexes

$$\chi : \bigoplus_k \mathcal{H}_n^{\otimes k} \to \bigoplus_k C_k(\mathcal{R}, \mathcal{R}^*).$$

As the morphism χ is injective, Φ is a cyclic cocycle.

4.7 Calculation of Φ

It is rather easy to calculate the pairing of the cyclic cocycle ϕ with elements of the K-theory of $C_c^\infty(P) \rtimes \Gamma$ coming from the Baum-Connes map. From this, it is not very hard to see which Pontryagin classes are involved in the formula. Further, Connes and Moscovici show that the secondary classes don't ultimately appear in the formula.

4.8 A variant of the Hopf algebra (cf. [23])

To construct the classifying map $H^*(W\,SO(n)) \to \mathrm{HC}^*_{\mathrm{per}}(\mathcal{A})$, we must start by replacing M by an open subset of $\mathbb{R}^n$. This step can be changed a little in the following manner.

To construct the Hopf algebra $\mathcal{H}_n$, we used the subgroup G_1 of diffeomorphisms of $\mathbb{R}^n$ which act freely and transitively on the frames. Such a group does not exist for a general manifold M. On the other hand, we can consider the groupoid $R \times R$, where R is the space of frames on M. This gives rise to a Hopf algebra $\mathcal{H}_M$ which is a quantum groupoid, rather than a quantum group. We can then construct a Hopf cyclic cohomology $\mathrm{HC}^*_{\mathrm{Hopf}}(\mathcal{H}_M)$, as well as a classifying map $\mathrm{HC}^*_{\mathrm{Hopf}}(\mathcal{H}_M) \to \mathcal{H}^*(\mathcal{A})$ and a cocycle $\Phi_M \in \mathrm{HC}^*_{\mathrm{Hopf}}(\mathcal{H}_M)$ with image ϕ.

Finally, it can be shown that the canonical morphism $\mathrm{HC}^*_{\mathrm{Hopf}}(\mathcal{H}_n) \to \mathrm{HC}^*_{\mathrm{Hopf}}(\mathcal{H}_M)$ is an isomorphism.

Remark.(cf. [4]) It is not in fact difficult to associate a "quantum groupoid" to a transitive action of a group G on a space X, along the lines discussed in 4.1.

References

1. M.F. Atiyah, *Global theory of elliptic operators*, Proc. Int. Conf. Functional Analysis and Related Topics. Univ. Tokyo Press (1970), 21–30.
2. S. Baaj et G. Skandalis, *Unitaires multiplicatifs et dualité pour les produits croisés de C^*-algèbres,* Ann. Sci. Éc. Norm. Sup. 4e série **26** (1993), 425–488.
3. S. Baaj et G. Skandalis, *Transformations pentagonales*, C. R. Acad. Sci. Paris Sér. I Math. **327**(7) (1998), 623–628.
4. S. Baaj et G. Skandalis, *in preparation*.
5. R. Beals and P. Greiner, *Calculus on Heisenberg manifolds,* Annals of Mathematics Studies **119**, Princeton University Press, Princeton, NJ (1988).
6. L.G. Brown, R.G. Douglas and P.A. Fillmore, *Extensions of C^*-algebras and K-homology,* Ann. of Math. **105** (1977), 265–324.
7. P. Cartier, *Homologie cyclique: rapport sur des travaux récents de Connes, Karoubi, Loday, Quillen...*, Sém. Bourbaki, 1983/84, Astérisque **121-122**, Soc. Math. France (1985), 123–146.
8. J. Cheeger, *Spectral geometry of singular Riemannian spaces,* J. Differential Geom. **18**(4) (1983), 575–657.
9. A. Connes, *Sur la théorie non commutative de l'intégration*, Lect. Notes in Math. **725** (1979), 19–143.
10. A. Connes, *Non commutative differential geometry. Chapter I: The Chern character in K-homology. Chapter II: De Rham homology and non commutative algebra,* Publ. Math. Inst. Hautes Études Sci. Publ. Math. No. **62** (1985), 257–360.
11. A. Connes, *Cyclic cohomology and the transverse fundamental class of a foliation,* Geometric methods in operator algebras (Kyoto, 1983), Longman Sci. Tech., Harlow, Pitman Res. Notes Math. Ser. **123** (1986), 52–144.

12. A. Connes, *Entire cyclic cohomology of Banach algebras and characters of θ-summable Fredholm modules*, K-Theory **1**(6) (1988), 519–548.
13. A. Connes, *Non commutative geometry*, Academic Press, Inc., San Diego, CA (1994).
14. A. Connes, *Geometry from the spectral point of view*, Letters in Mathematical Physics **34** (1995), 203–238.
15. A. Connes, *Trace formula in noncommutative geometry and the zeros of the Riemann zeta function*, Selecta Math. (N.S.) **5**(1) (1999), 29–106.
16. A. Connes and D. Kreimer, *Hopf algebras, renormalization and noncommutative geometry*, Comm. Math. Phys. **199**(1) (1998), 203–242.
17. A. Connes and D. Kreimer, *Renormalization in quantum field theory and the Riemann-Hilbert problem I: The Hopf algebra structure of graphs and the main theorem*, Comm. Math. Phys. **210**(1) (2000), 249–273.
18. A. Connes and D. Kreimer, *Renormalization in quantum field theory and the Riemann-Hilbert problem. II: The β-function, diffeomorphisms and the renormalization group*, Comm. Math. Phys. **216**(1) (2001), 215–241.
19. A. Connes and H. Moscovici, *The local index formula in noncommutative geometry*, Geom. Funct. Anal **5**(2) (1995), 174–243.
20. A. Connes and H. Moscovici, *Hopf algebras, cyclic cohomology and the transverse index theorem*, Comm. Math. Phys. **198**(1) (1998), 199–246.
21. A. Connes and H. Moscovici, *Cyclic cohomology and Hopf algebras*, Letters in Mathematical Physics **48** (1999), 97–108.
22. A. Connes and H. Moscovici, *Cyclic Cohomology and Hopf Algebra Symmetry*, Letters in Mathematical Physics **52** (2000), 1–28.
23. A. Connes and H. Moscovici, *Differentiable cyclic cohomology and Hopf algebraic structures in transverse geometry* (preprint 2001).
24. M. Crainic, *Cyclic Cohomology of Hopf algebras and a Noncommutative Chern-Weil Theory* (preprint 1998).
25. I.M. Gel'fand and D.B. Fuks, *Cohomologies of the Lie algebra of formal vector fields (Russian)*, Izv. Akad. Nauk SSSR Ser. Mat. **34** (1970), 322–337.
26. G. Godbillon, *Cohomologies d'algèbres de Lie de champs de vecteurs formels*, Sém. Bourbaki, 1971/72, exp. n° 421, Lect. Notes in Math. **383** (1974), 69–87.
27. J. Gracia-Bondía, J. Várilly and H. Figueroa, *Elements of noncommutative geometry*, Birkhäuser Advanced Texts: Basler Lehrbücher, Boston, MA (2001).
28. M. Hilsum et G. Skandalis, *Morphismes K-orientés d'espaces de feuilles et fonctorialité en théorie de Kasparov*, Ann. Sci. Éc. Norm. Sup. 4e série **20** (1987), 325–390.
29. A. Jaffe, A. Lesniewski and K. Osterwalder, *Quantum K-theory. I. The Chern character*, Comm. Math. Phys. **118**(1) (1988), 1–14.
30. G.I. Kac, *Extensions of groups to ring groups*, Math U.S.S.R. Sbornik **5** (1968), 451–474.
31. G.G. Kasparov, *The operator K-functor and extensions of C^*-algebras*, Math. USSR Izv. **16**(3) (1981), 513–572. *Translated from:* Izv. Akad. Nauk. S.S.S.R. Ser. Mat. **44** (1980), 571–636.
32. C. Kassel, *Le résidu non commutatif (d'après M. Wodzicki)*, Séminaire Bourbaki, vol. 1988/89, exposé n° 708, Astérisque **177-178** (1989), 199–229.
33. D. Kreimer, *On the Hopf algebra structure of perturbative quantum field theories*, Adv. Theor. Math. Phys. **2**(2) (1998), 303–334.
34. J. Kustermans and S. Vaes, *Locally compact quantum groups*, Ann. Sci. Éc. Norm. Sup. 4e série **33**(6) (2000), 837–934.

35. J-M. Lescure, *Triplets spectraux pour les pseudovariétés à singularité conique isolée*, C. R. Acad. Sci. Paris Sér. I Math. **327**(10) (1998), 871–874.
36. J-M. Lescure, *Triplets spectraux pour les pseudo-variétés avec singularité conique isolée*. PhD dissertation (preprint 1997).
37. S. Majid, *Hopf-von Neumann algebra bicrossproducts, Kac algebra bicrossproducts, and the classical Yang-Baxter equations*, J.F.A. **95**(2) (1991), 291–319.
38. R. Ponge, *Calcul hypoelliptique sur les variétés de Heisenberg, résidu non commutatif et géométrie pseudo-hermitienne*. PhD dissertation (preprint 2000).
39. R. T. Seeley, *Complex powers of an elliptic operator. Singular Integrals*, Proc. Sympos. Pure Math., Chicago, Ill. (1966) Amer. Math. Soc., Providence, R.I. (1967), 288–307.
40. M.E. Sweedler, *Hopf algebras*, Mathematics Lecture Note Series, W. A. Benjamin, Inc., New York (1969).
41. M. Takeuchi, *Matched pairs of groups and bismashed product of Hopf algebras*, Comm. Alg. **9** (1981), 841–882.
42. B.L. Tsygan, *Homology of matrix Lie algebras over rings and the Hochschild homology (in Russian)*, Uspekhi Mat. Nauk **38**, n° 2 (230) (1983), 217–218.
43. S. Vaes and L. Vainerman, *Extensions of locally compact quantum groups and the bicrossed product construction* (preprint 2001).
44. M. Wodzicki, *Noncommutative residue. I. Fundamentals*, K-theory, arithmetic and geometry (Moscow, 1984-1986), Lecture Notes in Math. **1289** (Springer, Berlin) (1987), 320-399.

Index

1_A 19
B 15
B' 16
$CC_\bullet(A)$ 74
$CC_\bullet^-(A)$ 74
$CC_\bullet^{\text{per}}(A)$ 74
$C_\bullet^\lambda(A)$ 74
$C_n^\lambda(A)$ 12
$C_\bullet(C^\bullet(A))$ 84
$D_0\{D_1,\ldots,D_m\}$ 80
$F^n X(T)$ 23
$HC_\bullet^-(A)$ 77
$HC_\bullet^{per}(A)$ 77
$HC_\bullet(A)$ 77
$HC_n(A,B)$ 19
$HC_n A$ 10
$HH_*(A)$ 10
$HP^*(A)$ 17
$HP_*(A)$ 17
$HP_*(A,B)$ 18
$H^\bullet(A)$ 80
$H^\bullet(A,A)$ 80
$H_*^{\text{loc}}(K,L)$ 58
L_D 83
$M_\infty(A)$ 6
N 14
P 14
Q 9
S 11, 17
S_D 83
$T\mathcal{A}$ 63
$X(A)$ 21
ΩA 13

$\Omega^1 A_\natural$ 21
$\alpha \oplus \beta$ 43
β 25
δ 25
κ 14
λ 9
$\mathcal{A}(a,b)$ 62
$\mathcal{A}[a,b]$ 62
$\natural$ 21
Sh 81
sh' 81
$\overline{\Omega}A$ 16
$\varepsilon_\mathcal{A}$ 44
$\widehat{\Omega}A$ 16
b 9, 14
b' 9, 87
ch 50
$ch(E)$ 33
$ch([e])$ 34
$ch([u])$ 34
d 14
i_D 82
$\mathcal{V}^\bullet(A)$ 93
$\mathcal{V}^\bullet(M)$ 93
$\mathbb{A}^t(M)$ 105
$\text{Calc}(M)$ 93
$\mathfrak{g}_A^\bullet$ 90

A_∞ algebra 84
A_∞ module 84
abstract differential forms 13
almost multiplicative 56
analytically nilpotent 56
analytically nilpotent curvature 56

Index

analytically quasi-free 57
associator 76

(B, b)-bicomplex 15
bicomplex 8
bivariant Chern character for
    C^*-algebras 60
bivariant Chern-Connes character 49
bivariant entire cyclic homology 52
bivariant periodic cyclic homology 18
bornological algebras 51
bornology 52
Bott map 45

calculus 93
canonical trace 103
chain homotopic 7
classifying map 44
complete 52
complete symbol 101
complex 7
complex powers 117
connecting morphism 88
connection 22
cyclic n-cocycle 13
cyclic bicomplex 9
cyclic cohomology 117
cyclic cohomology of Hopf algebras
    128
cyclic homology for Schatten ideals 38
cyclic object 96

de Rham cohomology 28
de Rham theory 38
deformation quantization 102
degree 7
differentiable homotopy 29
differential operator on differential
    forms 95
diffotopic 43, 62

entire cyclic cohomology 51
enveloping algebra of a Gerstenhaber
    algebra 95
excision 33

Fedosov product 24
filtration on $X(T)$ 23
Fredholm module 40

free product 63

general index theorem for deformations
    108
Gerstenhaber algebra 80
Gerstenhaber bracket 80
Goodwillie's theorem 27

H-unital 10, 11
Hochschild chain complex 74
Hochschild cochain complex 74
Hochschild homology 10
Hochschild operator 9
Hochschild-Kostant-Rosenberg theorem
    27
holomorphic function 98
Hom-complex 7
homology 7
homotopic 7
homotopy invariance for cyclic theory
    29
HP-equivalence 19

index theorem from symplectic
    deformations 75
index theorem of Fedosov 108
infinitesimal cohomology 28

JLO-cocycle 55

K-homology 118
Karoubi operator 14

Lie algebra homology 88
linearly split 44
local cyclic homology 59
local index formula 120
locally convex algebra 36
long exact sequence in homology 8

m-algebra 36, 61
matched pairs of groups 125
mixed complex 19
Morita context 32
Morita invariance 31

negative cyclic complex 74

p-summable cycles 118
p-summable Fredholm module 40
periodic cyclic cohomology 118

periodic cyclic complex   74
periodic cyclic homology   17
precalculus   93
principal symbol morphism   107
product on bivariant periodic cyclic
    theory   19
projective tensor product   61
pseudo-differential operator   101

quasi-free   22

reduced cyclic complex   87
Riemann-Roch theorem for D-modules
    109

SBI-sequence   11
Schatten ideal   65
shuffle product   81
simplicial object   96
small   52
smooth   27
smooth compact operator   64

smooth function   98
smooth subalgebra   59
smooth tensor algebra   62
spectral triple   120
supertrace   26
symplectic manifold   107

tensor algebra   6
θ-summable   51
Toeplitz algebra   65
total complex   8
trace density map   76
trace density morphism   107

unbounded Fredholm module   54
unitization   5

Wodzicki noncommutative residue   101
Wodzicki-Guillemin residue   117

X-complex   21